Makoto Yokoo, Takayuki Ito, Minjie Zhang, Juhnyoung Lee and Tokuro Matsuo (Eds.)

Electronic Commerce

Studies in Computational Intelligence, Volume 110

Editor-in-chief
Prof. Janusz Kacprzyk
Systems Research Institute
Polish Academy of Sciences
ul. Newelska 6
01-447 Warsaw
Poland
E-mail: kacprzyk@ibspan.waw.pl

Further volumes of this series can be found on our homepage: springer.com

Vol. 88. Tina Yu, David Davis, Cem Baydar and Rajkumar Roy (Eds.)
Evolutionary Computation in Practice, 2008
ISBN 978-3-540-75770-2

Vol. 89. Ito Takayuki, Hattori Hiromitsu, Zhang Minjie and Matsuo Tokuro (Eds.)
Rational, Robust, Secure, 2008
ISBN 978-3-540-76281-2

Vol. 90. Simone Marinai and Hiromichi Fujisawa (Eds.)
Machine Learning in Document Analysis and Recognition, 2008
ISBN 978-3-540-76279-9

Vol. 91. Horst Bunke, Kandel Abraham and Last Mark (Eds.)
Applied Pattern Recognition, 2008
ISBN 978-3-540-76830-2

Vol. 92. Ang Yang, Yin Shan and Lam Thu Bui (Eds.)
Success in Evolutionary Computation, 2008
ISBN 978-3-540-76285-0

Vol. 93. Manolis Wallace, Marios Angelides and Phivos Mylonas (Eds.)
Advances in Semantic Media Adaptation and Personalization, 2008
ISBN 978-3-540-76359-8

Vol. 94. Arpad Kelemen, Ajith Abraham and Yuehui Chen (Eds.)
Computational Intelligence in Bioinformatics, 2008
ISBN 978-3-540-76802-9

Vol. 95. Radu Dogaru
Systematic Design for Emergence in Cellular Nonlinear Networks, 2008
ISBN 978-3-540-76800-5

Vol. 96. Aboul-Ella Hassanien, Ajith Abraham and Janusz Kacprzyk (Eds.)
Computational Intelligence in Multimedia Processing: Recent Advances, 2008
ISBN 978-3-540-76826-5

Vol. 97. Gloria Phillips-Wren, Nikhil Ichalkaranje and Lakhmi C. Jain (Eds.)
Intelligent Decision Making: An AI-Based Approach, 2008
ISBN 978-3-540-76829-9

Vol. 98. Ashish Ghosh, Satchidananda Dehuri and Susmita Ghosh (Eds.)
Multi-Objective Evolutionary Algorithms for Knowledge Discovery from Databases, 2008
ISBN 978-3-540-77466-2

Vol. 99. George Meghabghab and Abraham Kandel
Search Engines, Link Analysis, and User's Web Behavior, 2008
ISBN 978-3-540-77468-6

Vol. 100. Anthony Brabazon and Michael O'Neill (Eds.)
Natural Computing in Computational Finance, 2008
ISBN 978-3-540-77476-1

Vol. 101. Michael Granitzer, Mathias Lux and Marc Spaniol (Eds.)
Multimedia Semantics - The Role of Metadata, 2008
ISBN 978-3-540-77472-3

Vol. 102. Carlos Cotta, Simeon Reich, Robert Schaefer and Antoni Ligeza (Eds.)
Knowledge-Driven Computing, 2008
ISBN 978-3-540-77474-7

Vol. 103. Devendra K. Chaturvedi
Soft Computing Techniques and its Applications in Electrical Engineering, 2008
ISBN 978-3-540-77480-8

Vol. 104. Maria Virvou and Lakhmi C. Jain (Eds.)
Intelligent Interactive Systems in Knowledge-Based Environment, 2008
ISBN 978-3-540-77470-9

Vol. 105. Wolfgang Guenthner
Enhancing Cognitive Assistance Systems with Inertial Measurement Units, 2008
ISBN 978-3-540-76996-5

Vol. 106. Jacqueline Jarvis, Dennis Jarvis, Ralph Rönnquist and Lakhmi C. Jain (Eds.)
Holonic Execution: A BDI Approach, 2008
ISBN 978-3-540-77478-5

Vol. 107. Margarita Sordo, Sachin Vaidya and Lakhmi C. Jain (Eds.)
Advanced Computational Intelligence Paradigms in Healthcare - 3, 2008
ISBN 978-3-540-77661-1

Vol. 108. Vito Trianni
Evolutionary Swarm Robotics, 2008
ISBN 978-3-540-77611-6

Vol. 109. Panagiotis Chountas, Ilias Petrounias and Janusz Kacprzyk (Eds.)
Intelligent Techniques and Tools for Novel System Architectures, 2008
ISBN 978-3-540-77621-5

Vol. 110. Makoto Yokoo, Takayuki Ito, Minjie Zhang, Juhnyoung Lee and Tokuro Matsuo (Eds.)
Electronic Commerce, 2008
ISBN 978-3-540-77808-0

Makoto Yokoo
Takayuki Ito
Minjie Zhang
Juhnyoung Lee
Tokuro Matsuo
(Eds.)

Electronic Commerce

Theory and Practice

With 78 Figures and 6 Tables

Prof. Makoto Yokoo
Graduate School of Information Science
and Electrical Engineering
Kyushu University, Hakozaki
6-10-1, Higashi ward
Fukuoka City, 812-8581, Japan
yokoo@is.kyushu-u.ac.jp

Prof. Takayuki Ito
Visiting Scholar
Sloan School of Management
Massachusetts Institute of Technology, USA
Visiting Scholar
Division of Engineering and Applied Science
Harvard University, USA
&
Nagoya Institute of Technology
Shikumi-college
Gokiso Showa-ku
Nagoya 466-8555, Japan
ito@eecs.harvard.edu
itota@ics.nitech.ac.jp

Prof. Minjie Zhang
Associate Professor
University of Wollongong
Wollongong, NSW 2522, Australia
minjie@uow.edu.au

Dr. Juhnyoung Lee
Research Staff Member
IBM T.J. Watson Research Center in
Hawthorne
19 Skyline Drive
Hawthorne, New York 10532, USA
jyl@us.ibm.com

Prof. Tokuro Matsuo
Graduate School of Science and Engineering
Yamagata University
4-3-16, Jonan
Yonezawa, 992-8510, Japan
matsuo@tokuro.net

ISBN 978-3-642-09656-3 e-ISBN 978-3-540-77809-7
DOI: 10.1007/978-3-540-77809-7

Studies in Computational Intelligence ISSN 1860-949X

Cover design: Deblik, Berlin, Germany

Printed on acid-free paper

9 8 7 6 5 4 3 2 1

springer.com

Preface

Electronic Commerce technologies have been studied widely in the field of Economics and Informatics. In recent years, there are multiple aspects of researches in electronic commerce including supply chain model, Internet auctions, Internet volume discounts, electronic bank and stock trading, and several others. The increase of Electronic Commerce research activities can be observed in a variety of the fields. The aim of this book is to encourage activities in this field, and to bring researchers with an interest in Electronic Commerce. This book is edited from some aspects of e-commerce researches including theoretical mechanism design of trading based on auctions, allocation mechanism based on negotiation among multi-agent, case-study and analysis of e-trading, data engineering issues in e-commerce, and so on. The reason why the many aspects are included is to show the outline of the newest research result of e-commerce research. We can collapse trends of the research to know them. Finally, we would like to express our sincere thanks to all authors for their hard work. This book would not have been possible without the valuable support and contributions of the cooperators.

Tokyo, December 1, 2007.

Makoto Yokoo
Takayuki Ito
Minjie Zhang
Junyoung Lee
Tokuro Matsuo

Contents

List of Contributors

Katsuhide Fujita
Department of Computer Science,
Nagoya Institute of Technology,
Gokiso, Showaku, Nagoya 466-8555,
Japan.
fujita@longwood.mta.nitech.ac.jp

Sangho Ha
School of Computer Science and Engineering, Soonchunhyang University,
Asan-si, Choongnam, 336-745, Korea.
jshur@sch.ac.kr

Hiromitsu Hattori
Graduate School of Informatics, Kyoto University, Japan.
hatto@i.kyoto-u.ac.jp

Jung-Su Hur
School of Computer Science and Engineering, Soonchunhyang University,
Asan-si, Choongnam, 336-745, Korea.
hsh@sch.ac.kr

Takayuki Ito
Techno-Business School, Nagoya
Institute of Technology, Gokiso,
Showaku, Nagoya 466-8555, Japan.
ito.takayuki@nitech.ac.jp

Ok-ran Jeong
School of Computer Science and
Engineering, Seoul National University,
Seoul 151-742 Korea.
orjeong@europa.snu.ac.kr

Mark Klein
Center for Collective Intelligence,
Massachusetts Institute of Technology,
USA.
m_klein@mit.edu

Masao Kobayashi
Department of Computer Science,Nagoya Institute of Technology,
Gokiso, Showaku, Nagoya 466-8555,
Japan.
kobayashi@longwood.mta.nitech.ac.jp

Hyunja Lee
Department of Computer Science,
Sookmyung Women's University,
Seoul 140-742, Korea.
hyunjalee@sookmyung.ac.kr

Sang-goo Lee
School of Computer Science and
Engineering, Seoul National University,
Seoul 151-742 Korea.
sglee@europa.snu.ac.kr

Taehee Lee
School of Computer Science and Engineering, Seoul National University, Seoul 151-742 Korea.
thlee@europa.snu.ac.kr

Shigeo Matsubara
NTT Communication Science Laboratories, NTT Corporation,
2-4 Hikaridai, Seika-cho, Soraku-gun, Kyoto 619-0237, Japan
matsubara@cslab.kecl.ntt.co.jp

Masafumi Matsuda
NTT Communication Science Laboratories, NTT Corporation,
2-4 Hikaridai, Seika-cho, Soraku-gun, Kyoto 619-0237, Japan
masafumi@cslab.kecl.ntt.co.jp

Tokuro Mastuo
Yamagata University, 4-3-16, Jonan, Yonezawa, Yamagata, 992-8510, Japan.
matsuo@yz.yamagata-u.ac.jp

Angel Merono-Cerdan
Dpto. de Organizacion de Empresas y Finanzas, Universidad de Murcia, Campus de Espinardo,
30.100 Murcia, Spain.
angelmer@um.es

Yoshihito Saito
Yamagata University, 4-3-16, Jonan, Yonezawa, Yamagata, 992-8510, Japan.
saito2007@e-activity.org

Junho Shim
Department of Computer Science, Sookmyung
Women's University, Seoul 140-742, Korea. jshim@sookmyung.ac.kr

Toramatsu Shintani
Graduate School of Engineering, Nagoya Institute of Technology, Gokiso, Nagoya, 466-8555, Japan.
matsuo@yz.yamagata-u.ac.jp

Pedro Soto-Acosta
Dpto. de Organizacion de Empresas y Finanzas, Universidad de Murcia, Campus de Espinardo, 30.100
Murcia, Spain. psoto@um.es

Satoshi Takahashi
Yamagata University, 4-3-16, Jonan, Yonezawa, Yamagata, 992-8510, Japan.
takahashi2007@e-activity.org

TakahiroWatanabe
Faculty of Urban Liberal Arts, Tokyo Metropolitan University, 1-1 Minami-Osaka, Hachioji, Tokyo
192-0397, Japan.
forward0@nabenavi.net

Takehiko Yamato
Department of Value and Decision Science,
and Department of Social Engineering, Graduate School of Decision
Science and Technology, Tokyo Institute of Technology, 2-12-1
Ookayama, Meguro-ku, Tokyo
152-8552, Japan.

Product Ontology and OWL Correspondence

H. Lee and J. Shim

Department of Computer Science,
Sookmyung Women's University,
Seoul 140-742, Korea,
{hyunjalee, jshim}@sookmyung.ac.kr

Summary. Product information is a core component in an e-commerce application. Semantically enriched and precise product information may enhance the quality and effectiveness of business transactions. A recent approach is to employ the ontology to model the product information. A basis is to model a formal product ontology which can be also applicable in practice. In addition, it may be beneficial to transform or publish the product ontology in a standard ontology language. A most well-known ontology language is OWL. In this paper, we present how each modeling construct of product ontology can be translated in OWL language. We take into account of the expressiveness and complexity provided by OWL for the translation of each modeling construct, along with its practical usage in the domain.

1 Introduction

Product information includes various types of information such as pricing, features, or terms about the goods and services. Having product information precisely and clearly defined is important in e-commerce area where product information may be searched and navigated by an application program, and also electronically interchanged between business partners. Recently, ontology has been applied to bring these features to product information [3][6][10][11][12]. Product ontology requires specifying a conceptualization of product information in terms of classes, properties, relationships, and constraints. From a project to building an operational product ontology database, we have learned what concepts, in terms of classes, properties, relationships, and constraints, may be fundamental to represent the domain [12]. In our preliminary work [13], we introduced a modeling approach which formally represents product ontology. In this work, we represented each modeling construct in description logics which provide a theoretical core for most of the current ontology languages.

A knowledgebase represented in a standard and machine-operational language may role as a basis to utilize a so-called ontology-engineering technique [7][20]. It may be possible and therefore desirable to make a knowledgebase loosely-coupled from the application codes, to develop knowledge bases independent and interoperable each other, and to automate the reasoning facility provided by inference engines

H. Lee and J. Shim: *Product Ontology and OWL Correspondence*, Studies in Computational Intelligence (SCI) **110**, 1–14 (2008)
www.springerlink.com

[18]. OWL (Web Ontology Language) [19] has been positioned as a most widely used ontology language.

In this paper, we present how to represent a product ontology model in OWL. OWL has three different versions Lite, DL, and full; according to its language syntax and thereby different complexity as well. We employ OWL-DL to represent product ontology. However, it still requires considering the employed syntax in terms of its expressiveness and complexity, since some are little usage in practice yet require high complexity to process.

The rest of this paper is structured as follows. Section 2 briefly introduces the related works and our product ontology model. In Section 3, we present a basic OWL correspondence to the elemental modeling constructs. Constraints should be selectively considered accordingly to the type of semantic relationships, which is explained in Section 4. And finally, we provide the related work and future research direction in Section 5.

2 Related Works

2.1 e-Catalog and Product Ontology

Researches in recent years show that applying ontology to e-commerce would bring benefits such as to solving the interoperability problems between different e-commerce systems[1][11][16][23]. Especially, e-Catalog, which is a key component in e-commerce systems, seems to be the most adequate domain within e-commerce where ontology can play an important role.

There have been papers expressed the early ideas on applying ontology into e-Catalog [3][17]. Fensel, et al [3] present the issues of B2B integration, focusing on product information. They list the difficult aspects of building, maintaining, and integrating product information, and propose that ontological approach may be the answer. Obrst, et al [17] proposes to use cross industry standard classifications such as UNSPSC [22] and eCl@ss [2] as the upper ontology and industry specific classifications as lower ontology, thus achieving the generality and specificity. These two early works are representative of a group of works focusing mainly on classification standards as the shape of ontology for product information. Classification hierarchies are essential part of product information semantics but make up only one piece of the picture. Also, their proposals are still in an abstract level.

An interesting effort is presented in [15]. They try to utilize the ISO standard for product library (PLIB) to model product ontologies. However, their view of product ontology is limited to classification hierarchy and the PLIB standard provides a set of meta data specifications for such hierarchies. Hepp [5], Kim, et al [8], and Lee [10] are important works that emphasize the importance of attributes in product information management. Hepp [5] evaluates the quality of product classification standards based a number of factors including the quality of their attribute lists. In [8], a data model for classification hierarchies is presented where specification of attributes and semantics is a requirement. In [10], it is pointed out that a classification hierarchy is a

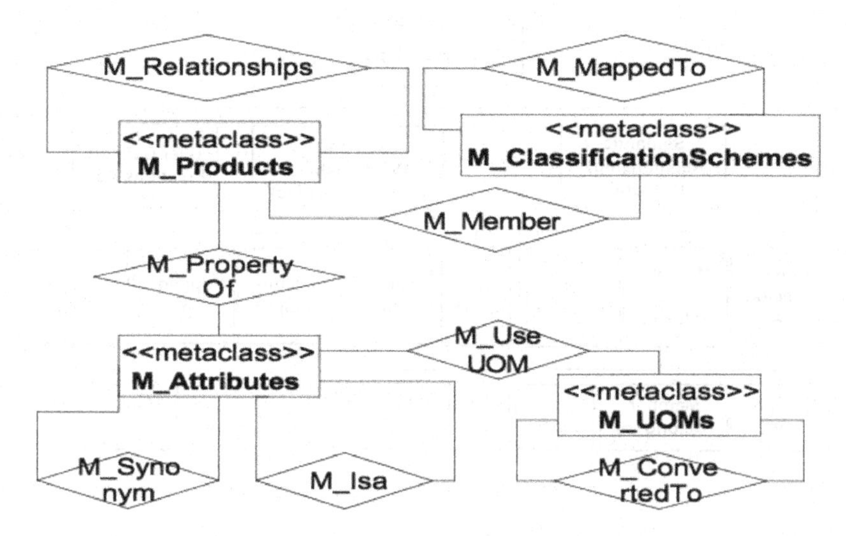

Fig. 1. Concepts and relationships in the meta-level product ontology

representation of just one of many views over the set of products. A product's identity and property does not depend on how the product is classified. Product database design issues and guidelines are presented, where the focus is on properties (attributes) rather than on classification hierarchies.

2.2 Product Ontology Model

For modeling product ontology, we need to investigate the key concepts and their relationships of the domain. From a real project [12] that we participated in, products, classification scheme, attributes, and UOMs are determined as the key concepts. Since an ontology model should be apt to the different points of views, we employ a meta-modeling approach to produce an extensible and flexible product ontology model. As shown in Figure 1, those concepts aforementioned are modeled as meta-concepts. The products, the most important concept, are for the goods or services. The classification schemes and the attributes are used for the classifications and descriptions of products, respectively. Then the UOMs, unit-of-measures, are associated with the attributes. The semantic relationships we consider include those from general domain [21]; such as class inclusion (isa), meronymic inclusion (component, substance, and member), attribution, and synonym. In addition, product domain specific relationships such as substitute, complement, purchase-set, mapped-to are also considered (Figure 2).

For examples, a LCD monitor is a substitute of a CRT monitor and vice-versa in that each may serve as a replacement of the other. A complement relationship represents that one may be added to another in order to complete something. For example, an antiglare filter is complement to a monitor. We may also see that such products as a monitor, OS, and mouse are purchased with a personal computer. This

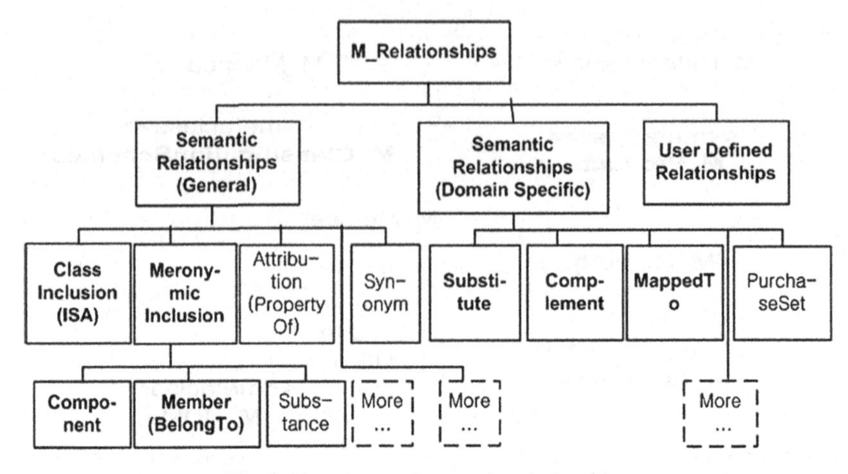

Fig. 2. Taxonomy of semantic relationships

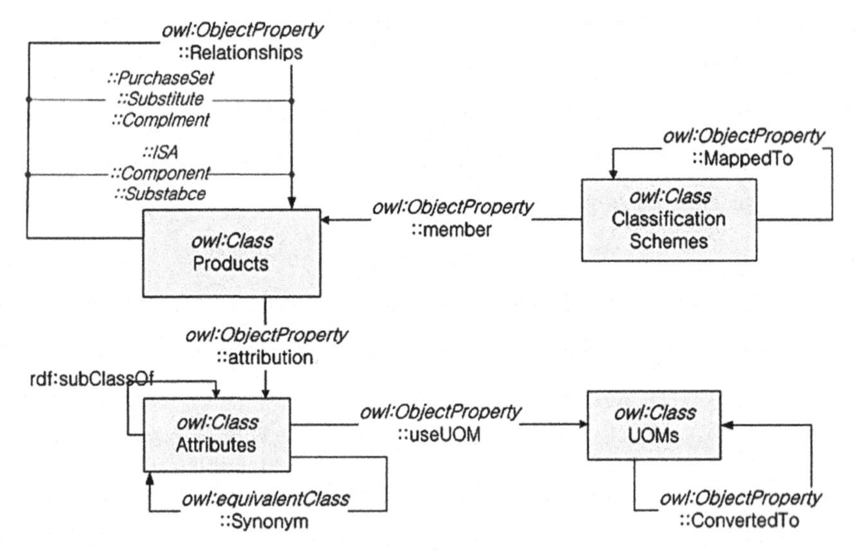

Fig. 3. Basic OWL correspondence in meta-level

is represented as purchase-set. And finally, the mapped-to relationship is to assign a product into a specific class code within a classification scheme, or to map a class code from a classification scheme to different schemes.

In [13], we illustrate that concepts and semantic relationships can be represented by the SHIQ(d) description language, which is reasonably practical with regard to its language expressiveness and algorithmic complexity [4]. Note that OWL-DL is a OWL version that corresponds to the description language. In the following section, we present how our product ontology model can be translated into OWL versions.

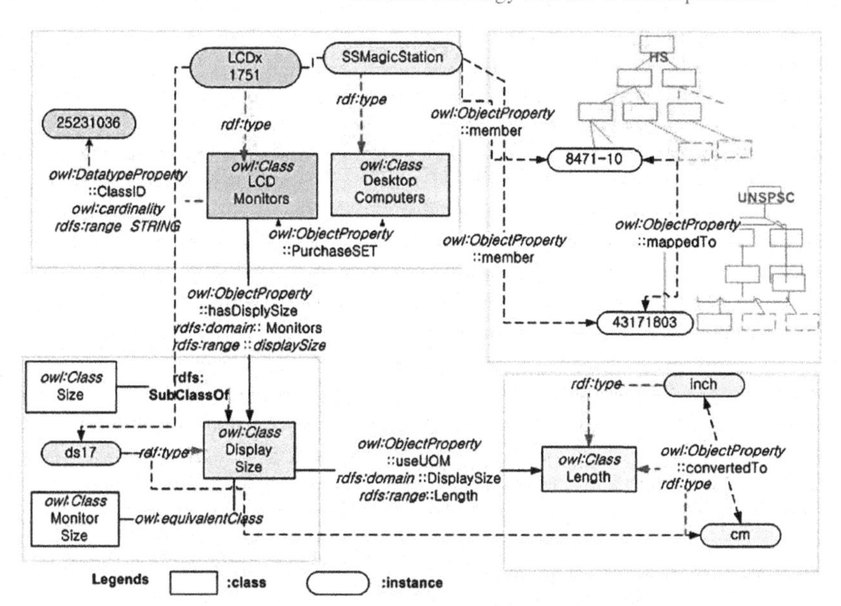

Fig. 4. A comprehensive OWL representation for computer-related products

3 Basic OWL Correspondences

The Figure 3 shows a basic mapping from our product meta-model (Figure 1) to an OWL representation. Note that we rename PropertyOf(Attribution) relationship between Products and Attributes with attribution to avoid confusion with the property expression in OWL.

Basically, concepts can be represented by using owl:Class, and relationships by owl:ObjectProperty or owl:DatatypeProperty, in general. The key concepts in the meta-class level may be correspondent to owl:Class. Most of relationships between concepts such as member, attribution, useUOM, convertedTO, mappedTo, and relationships including purchaseSET, substitute, complement, component, etc. may be correspondent to owl:ObjectProperty, and isa relationship between Products or between Attributes correspondent to rdf:subClassOf, and synonym relationship may be converted to owl:equivalentClass.

Note that in OWL, the domain of a property limits the individuals to which the property can be applied and the range of a property limits the individuals or values of which property may have. The datatype property is used for relationship which exists between a class instance and a data value, while the object property is used for relationships between instances. Relationships may have additional property restrictions or property characteristics (Section 4).

Figure 4 is a more comprehensive example to illustrate how the products such as desktop computer or LCD monitor as a member of the classification system, and their instance with attribute size relate each other in OWL. It also represents the relationships between the size of monitor and its associated UOM. In the figure, a class is denoted by a rectangle and an instance is denoted by an ellipse. The in and out going edge of an arc illustrates the domain and the range of a property respectively.

As shown in the figure, desktop computers are a member of 8471-10 and of 43171803 commodities in HS and UNSPSC classification system, respectively. This member relationship may be represented using the owl:ObjectProperty::member object property with rdfs:domain::ClassificationScheme and rdfs:range::Product restrictions. LCD Monitors class has ClassID attribute and it has a string data type value of "25231036" as range. Then it is represented by owl:DatatypeProperty with a restriction of rdfs:range::STRING. ClassID is only one value per each class, and therefore it is restricted to cardinality using owl:cardinality.

Desktop computers and LCD Monitors are purchased together each other. This is represented using owl:ObjectProperty::purchaseSET with an additional restriction with owl:SymmetricProperty to denote the symmetric property of the purchaseSET property. Additionally, if the domain of purchaseSET is limited to Desktop computers, then the range should be limited by adding owl:someValuesFrom restriction.

LCDx1751QD or SAMSUNG MagicStation is an individual of LCD Monitors class or Desktop computers class, respectively. The individual products are represented by declaring it to be a member of a class like

$$< LCDMonitorsrdf : ID = "LCDx1751QD" >$$

and

$$< DesktopComputerrdf : ID = "SAMSUNGMgicStation" >$$

respectively. As following, rdf:type is another expression of a RDF property that ties an individual to a class of which it is a member.

```
<owl:Thing rdf:about="# LCDx1751QD ">
   <rdf:type rdf:resource="# LCDMonitors "/>
</owl:Thing>
```

A LCD Monitor has the Size attribute, and this is represented using owl:ObjectProperty::hasDisplaySize of which domain and range are specified to LCD Monitors and DisplaySize. DisplaySize is a sub type of Size and also synonym to MonitorSize, which is represented using rdfs:SubClassOf and owl:EquivalentClass, respectively. The DisplaySize attribute is associated with the Length UOM, of which instances include inch and cm. Similar to the previous examples, these may be represented using owl:ObjectPropertyOf::useUOM and rdf:type. Note that inch and cm UOMs may be converted each other, i.e., 1inch = 2.54cm. The conversion equation can be represented and processed by a knowledge representation language, while OWL is yet incapable of providing linear polynomial equations [4].

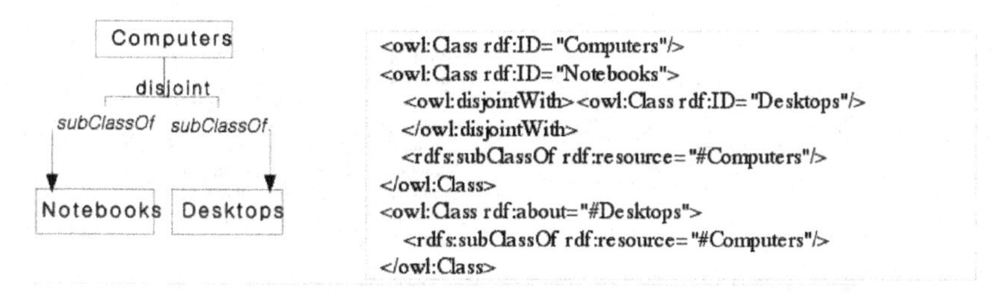

Fig. 5. OWL representation for the inclusion relationship

4 Property Restrictions for Semantic Relationships

As mentioned in the previous section, although relationships are corresponded to owl:ObjectProperty, more restrictions should be selectively added to each object property to convey precise semantics. Those restrictions include owl:Transitive-Property, owl:SymmetricProperty, owl:inverseOf, and owl:FunctionalProperty, and owl:some ValuesFrom and owl:allValuesFrom to further constrain the range of a property in specific contexts as well as domain and range of property. In Section 4.1, we first explain what restrictions are required for each type of semantic relations in general domain, and then in Section 4.2 for product domain specific ones.

4.1 Semantic Relationships in General Domain

General semantic relationships include inclusion, meronymic inclusion, attribution, and synonym. A inclusion relationship can be represented by using owl:Class and rdfs:subClassOf. A inclusion relationship may have additional constraints such as "disjoint". For example, in Figure 5, notebooks and desktops are the subclasses of computers and they are disjoint each other, which is represented using owl:disjointWith.

Second, let us consider a meronymic inclusion case. In Figure 6, DC spindle motor is a component (or part) of HDD(hard disk) product which is also component of computer products. A component (or part) relationship, in general, is not always transitive in that it usually contains both aggregational and functional semantics; being a part functional for its whole does not necessarily mean that the part again plays functional for another object which is composed of the whole [21].

However in practice, people often do not clarify the precise semantic of a component relationship that they use. For example, a query to find all hardware classes which contain DC Spindle as its component is not clear whether it is meant to inquire to search for any hardware having DC Spindle as its direct part or as its both direct and indirect part (being contained within another product). In order to handle indirect part-whole relationship, we need the transitive property, i.e., x part of ẏ & y part of x part of ż. This can be represented in OWL using owl:TransitiveProperty. And if HDD is a component(part) of Computer, then Computer is the composed(whole) of HDD. This may be represented using owl:inverseOf. In addition, the value restriction owl:someValuesFrom should be applied to the component property since Computers

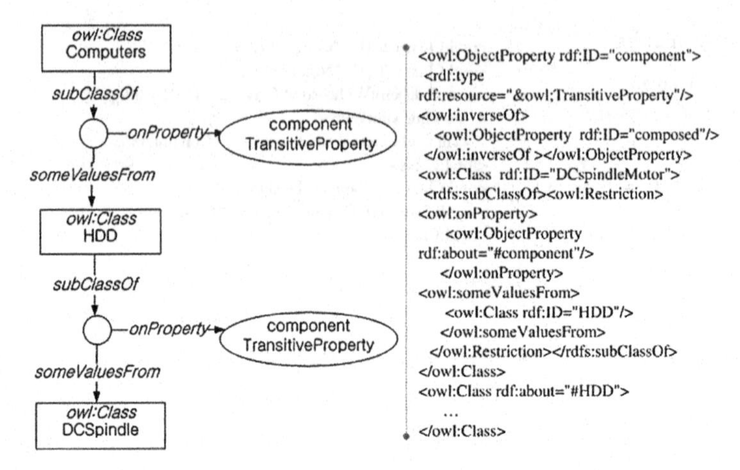

Fig. 6. Restrictions on the component relationship

may have not only HDD but also other parts such as CPU, graphic card, RAM and etc.

Instead of having a component relationship, we may have a domain specific component relationship between a part and a whole and declare it as a subproperty (owl:subPropertyOf) of the component relationship. Note that the subproperty in OWL corresponds to the role inclusion in logic representation. For example in Figure 6, we may define a sub property of component, computerComponent, of which domain and range is restricted to Computers and HDD respectively, and use it rather than the component property. Then we need to add the followings, to declare that computerComponent is a role inclusion of component.

```
<owl:ObjectProperty rdf:about="\#computerComponent">
<rdfs:subPropertyOf>
    <owl:ObjectProperty rdf:about="\#component">
</rdfs:subPropertyOf>
</owl:ObjectProperty>
```

Readers should note that if we use a domain specific component subproperty we should use the specific component in a query to retrieve the components in that domain. This principle can be applied to other types of semantic relationships.

An attribution relationship may be represented using owl:DatatypeProperty in accordance with the xml schema datatypes. In Figure 7, ClassID, one of the LCDMonitors attributes, is represented by its data value with string. ClassID is only one value per each class, and therefore it is restricted to cardinality using owl:cardinality, i.e, $< owl : cadinalityrdf : datatype = "\&xsd; string > 1 < /owl : cadinality >$.

An attribution relationship which exists between Products and Attributes may be represented by using owl:objectProperty in that a product attribute may be managed

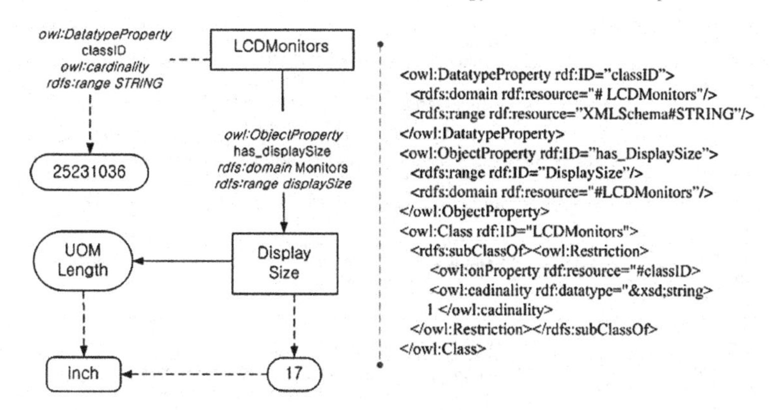

Fig. 7. Restrictions on the attribution relationship

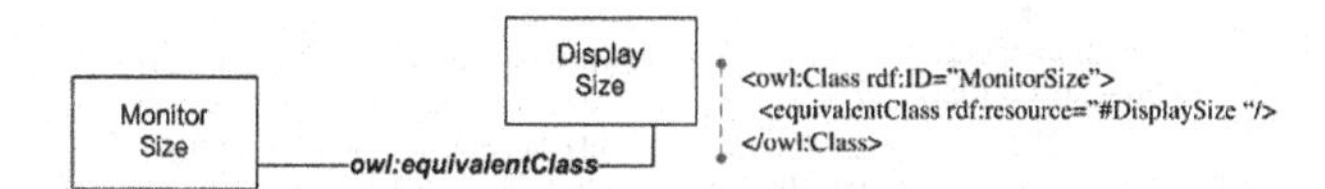

Fig. 8. Restrictions on synonym relationship

as a class. In Figure 7, for example, we have the has_displaySize relationship between LCDMonitors and its attribute, DisplaySize. The has_displaySize should be represented using an object property expression rather than datatype property. Note that 17 is not a data value but an individual of DisplaySize class and the DisplaySize attribute uses UOM Length inch.

And finally, a synonym relationship is represented by using owl:equivalentClass. In Figure 8, MonitorSize is an alias to DisplaySize, and therefore they are equivalent classes each other.

4.2 Domain Specific Semantic Relationships

While the semantic relationships presented in the previous section may be observed in a general domain, those presented in this section are particularly conceivable relationships in product information domain. They are the substitute, supplement (or complement), purchase-set, and mapped-to relationships as shown in Figure 2 and 3.

Let us consider an example that there are supplement(complement) relationships between Antiglare filter and Monitor or between Toner and Laser Printer (Figure 9). Let us assume that an individual antiglare filter product may serve as a supplement to any monitor product while only specific type of toner product may serve a laser printer. For example, only TonerHP-2420 but other toner products may work for Laserjet2420. In this case, owl:Objectproperty:supplement with owl:allValuesFrom may be enough to represent the supplement relationship between Antiglare filter and

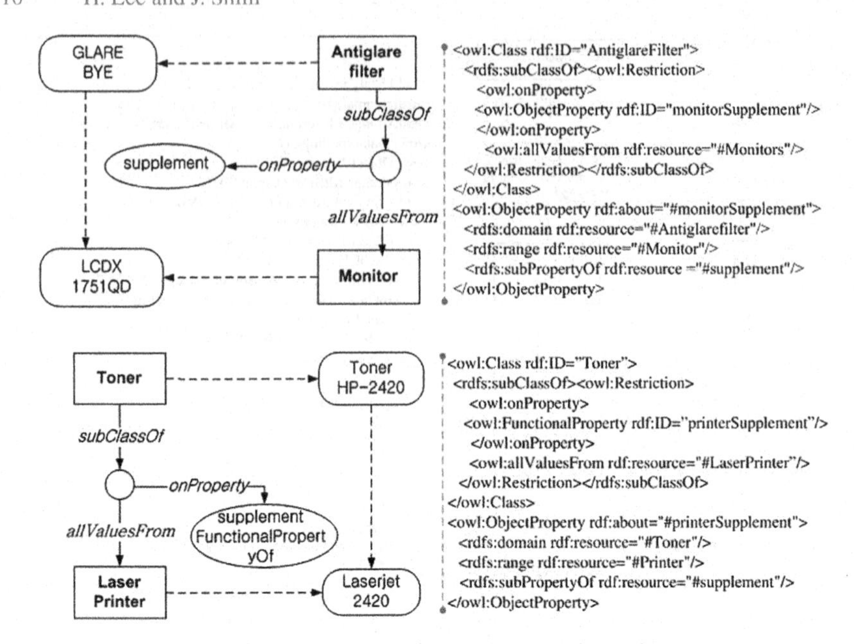

Fig. 9. Restrictions on the supplement relationship

Monitor product classes. Whereas we should add owl:FuctionalProperty restriction to the supplement relationship between TonerHP-2420 and Laserjet2420.

In addition, supplement relationships should be specified along what product relevant to domain or range. In the above example, property constraints such as functional property, may be applied or not dependently on the domain or range of property. Two supplement relationships in the figure are represented in different ways each other. It may be specified as monitorSupplement and printerSupplement by subtypes of supplement relationships with OWL expression as rdfs:subPropertyOf. The monitorSupplement and printerSupplement are restricted to owl:allValuesFrom:: Monitors and owl:allValuesFrom::Printers, respectively.

A substitute relationship means that one may role as a replacement of the other. For example, a pencil is a substitute of a mechanical pencil. If a LCD monitor is a substitute of a CRT monitor, then a CRT monitor is a substitute of a LCD monitor, vice versa. The substitute relationships has the symmetric property, i.e., x substitute y _ y substitute x as well as the transitive property. For example, if sugar is a substitute of aspartame, and aspartame is a substitute of saccharin, then sugar may also be a substitute of saccharin (Figure 10). Then it may be represented in OWL as owl:ObjectProperty::substitute with the restriction of owl:SymmerticProperty, owl:TransitiveProperty.

A Substitute relationship with symmetric characteristic has the same domain and ranges each other. Along with domain or range scope, the relationship may be constrained as rdfs:subPropertyOf and owl:someValuesFrom. In Figure 10, a sub-

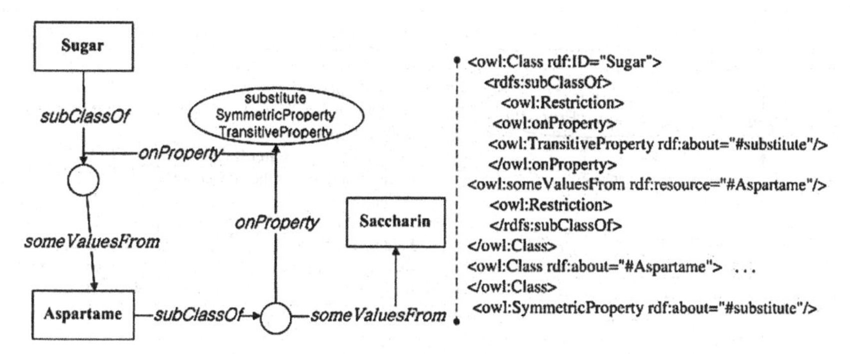

Fig. 10. Restrictions on the substitute relationship

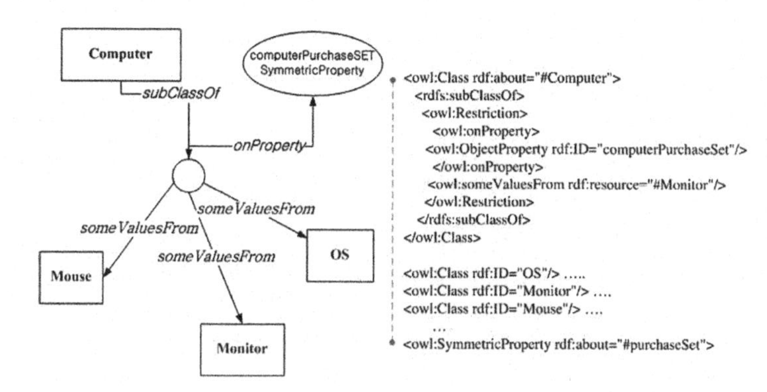

Fig. 11. Restrictions on the purchaseSET relationship

stitute relationship is illustrated among Sugar, Aspartame, and Saccharin. Sugar is substituted for Aspartame, while Aspartame is substituted for Saccharin. Since these substitution relationships contain both symmetric and transitive characteristics, Saccharin may be also substituted for Sugar, and vice versa.

And next, let us consider the purchaseSET relationship (Figure 11). A purchaseSET relationship also has the symmetric characteristic. The figure illustrates that a computer product may be purchased together with monitor, OS(operating system software), and mouse products. The domain of the purchaseSET property is Computer and its range may include Monitor, OS, and Mouse. In this case, we may employ computerPurchaseSET as a sub property of purchaseSET.

Finally, we can consider the mappedTo relationship (Figure 12). The mappedTO relationship exists between classification schemes, while any other domain specific relationships described above, are dealt with between products. Namely both the domain and range are assigned to classification schemes. It is also symmetric and transitive, which is represented by using owl:SymmetricProperty and owl:Transitive Property, respectively. Figure 12 illustrates that the HS classification scheme can be mapped to UNSPSC, and vice versa each other. Having a mappedTO relationship,

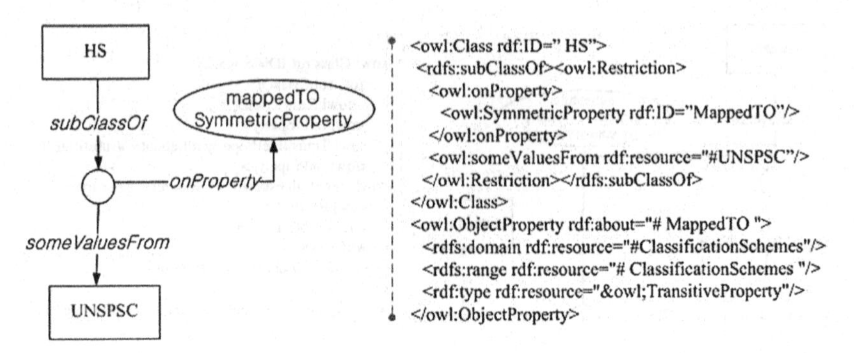

Fig. 12. Restrictions on the mappedTo relationship

we may then a mapping between individual codes in two different classification schemes. For example, in Figure 4 in Section 3, 8471-10 in HS is mapped to 43171803 in UNSPSC.

5 Conclusion

There have been studies in multiple domains for building an OWL-based ontology to represent the semantic knowledge of the domain. The domains include biology, life-science, medicine, business, and e-commerce [20]. Our work presented in the paper may not be a unique work in product ontology. In [3], they list the difficult aspects of building, maintaining, and integrating product information, and propose that ontological approach may be the answer. In [6], they focus on OWL derivation for industry standard taxonomy and classify concepts to capture the semantics of those standards.

It is important to investigate what semantic concepts and relationships, not only limited to the taxonomy of product classes, are desirable for the domain and to capture them in the model. In addition, the model should be formally represented to convey the precise semantics. A most popular way to formally describe a ontology may be using OWL language. Our contribution is to present an OWL representation scheme for product ontology of which model reflects a fundamental set of concepts, relationship, and constraints in the product information domain.

Another important issue in product ontology is not only to provide a formal model itself; but to implement a product ontology model so that applications may run in large scale and benefit the underlying ontology [11][12][14]. In [14], building a pure OWL knowledgebase for product ontology is considered little practical due to the lack of a robust engine to manage a large knowledgebase with a business demanding performance. Instead they propose to build a product ontology database using a commercial object-relational DBMS and also to provide a feature to export relational tables into an OWL knowledgebase. The scheme explained in this paper role as a theoretical background of the work.

Acknowledgement. This research was supported by the Sookmyung Women's University Research Grants 2007.

References

1. Z. Cui, D. Jones, and P. O'Brien, Semantic B2B Integration: Issues in Ontology-based Approaches, SIGMOD Record, Vol. 31, No. 1, ACM Press, 2002.
2. eCl@ss - New Standardized Material and Service Classification, http://www.eclass-online.com/, Cologne Institute for Business Research.
3. D. Fensel, Y. Ding, B. Omelayenko, E. Schulten, G. Botquin, M. Brown, and A. Flet, Product Data Integration in B2B E-Commerce, IEEE Intelligent Systems, 2001.
4. V. Haarslev and R. Moller, Description Logic Systems with Concrete Domains: Applications for the Semantic Web, 10th International Workshop on Knowledge Representation meets Databases (KRDB 2003), CEUR Workshop Proceedings, 2003.
5. M. Hepp, Measuring the Quality of Descriptive Languages for Products and Services, in E-Business, Standardisierung und Integration, F. D. Dorloff, et al(Editors), Tagungsband zur Multikonferenz Wirtschaftsinformatik, 2004.
6. M. Hepp, A Methodology for Deriving OWL Ontologies from Products and Services Categorization Standards, Proceeding of 13th European Conference on Information Systems (ECIS), 2005.
7. V. Kashyap and A. Borgida, Representing the UMLS Semantic Network Using OWL, International Semantic Web Conference, Springer-Verlag, 2003.
8. D. Kim, J. Kim, and S. Lee, Catalog Integration for Electronic Commerce through Category-Hierarchy Merging Technique, 12th International Workshop on Research Issues on Data Engineering (RIDE2002), IEEE Society, 2002.
9. K. Kim, M. Tark, H. Lee, J. Shim, J. Lee, and S. Lee, PROMOD: A Modeling Tool for Product ontology, 2nd International Workshop on Data Engineering Issues in E-Commerce and Services (DEECS 2006), Springer-Verlag, 2006.
10. S. Lee, Design & Implementation of an e-Catalog Management System, Tutorial at 9th International Conference on Database Systems for Advances Applications (DASFAA 2004), Springer-Verlag, 2004.
11. J. Lee and R. Goodwin, Ontology Management for Large-Scale E-Commerce Applications, Electronic Commerce Research and Applications, Vol. 5, No. 1, Elsevier, 2006.
12. I. Lee, S. Lee, T. Lee, S.-g. Lee, D. Kim, J. Chun, H. Lee, and J. Shim, Practical Issues for Building a Product Ontology System, International Workshop on Data Engineering Issues in E-Commerce (DEEC2005), IEEE Society, 2005.
13. H. Lee, J. Shim, and D. Kim, Ontological Modeling of e-Catalogs using EER and Description Logic, International Workshop on Data Engineering Issues in E-Commerce (DEEC2005), IEEE Society, 2005.
14. T. Lee, J. Shim, H. Lee, and S.-g. Lee, Database Modeling for Operational Product Ontology, Technical Report, Center for e-Business Technology, 2006 (submitted to Journal of Data Semantics).
15. J. Leukel, Standardization of Product Ontologies in B2B Relationships, On the Role of ISO 13584, Proceeding of 10th Americas Conference on Information Systems, 2004.
16. J. Leukel, V. Schmitz, and F. Dorloff, A Modeling Approach for Product Classification Systems, 13th International Conference on Database and Expert Systems Applications (DEXA 2002), Springer-Verlag, 2002.

17. L. Obrst, R.E. Wray, H. Liu, Ontological Engineering for B2B E-Commerce, International Conference on Formal Ontology in Information Systems (FOIS'01), ACM Press, 2001.
18. J. Pollock, Using the W3C Standard OWL in Adaptive Business Solutions, WWW2004 Conference, http://www.w3.org/2001/sw/EO/talks, 2004.
19. M.K. Smith, C. Welty, and D.L. McGuinness, OWL Web Ontology Language Guide, W3C Recommendation, http://www.w3c.org/TR/owl-guide/, 2004.
20. S. Staab and R. Studer (Eds.), Handbook on Ontologies. International Handbooks on Information Systems, Springer-Verlag, 2004.
21. V. C. Storey, Understanding Semantic Relationships, VLDB Journal, Vol. 2, VLDB Endowment, 1993.
22. UNSPSC, United Nations Standard Products and Services Code, http://www.unspsc.org/, UNDP.
23. G. Yan, W. K. Ng, and E. Lim, Product Schema integration for Electronic Commerce - A Synonym Comparison Approach, IEEE Transaction on Knowledge and Data Engineering, Vol. 14, No. 3, IEEE Society, 2002.

Optimizing the Pre-Processing Phase of Automatic e-Document Classification

Ok-ran Jeong, Taehee Lee, and Sang-goo Lee

School of Computer Science and Engineering,
Seoul National University,
Seoul 151-742 Korea
{orjeong, thlee, sglee}@europa.snu.ac.kr

Summary. Electronic documents such as e-catalogs, e-mails, and Web documents have their own distinct characteristics that can be utilized in search and classification. They are structured, noisy, and, in some cases, related to each other. We analyze the characteristics of three major types of e-documents - e-catalogs, e-mails, and Web documents - and propose methods for optimizing automatic classification of such documents. Our Proposal is to improve the pre-processing phase of automatic e-document classification by considering the document-specific characteristics, while continuing to exploit the state of the art classification algorithms. We expect that the pre-processing techniques can be applied to other types of e-documents that exhibit similar characteristics.

1 Introduction

Automatic e-document classification is challenging due to the diversity of features. Nonetheless, research has made headway for major document types, driven by the need to reduce work load required for document classification. Even when the same algorithm is used, it is possible to increase the accuracy of classification through effective pre-processing. The document classification process is largely divided into preprocessing stage, feature extraction stage and document classification stage. However, this research classifies $pre-processing$ stage and feature extraction into one preprocessing stage. By pre-processing, we refer to all work that must be processed prior to the efficient classification of documents. For example, the elimination of tags such as stop words or unnecessary terms is part of the pre-processing parameters.

Although pre-processing is essential for classifying any type of e-documents, it is even more important for three documents that are particularly amenable to pre-processing; e-catalogs, e-mails, and Web documents. E-catalogs are composed of short, product-oriented words. Pre-processing for product unit classification is very helpful. E-mails contain numerous non-standard words. Often, titles contain important clues for the contents, and pre-processing is very helpful. Web documents contain HTML tags, and are linked through hyper links.

O. Jeong et al.: *Optimizing the Pre-Processing Phase of Automatic e-Document Classification*, Studies in Computational Intelligence (SCI) **110**, 15–26 (2008)
www.springerlink.com

After the pre-processing, actual classification algorithm is applied to classify documents. The most often used methods for classifying documents are rule based approach that use rules derived from learning to classify documents [1].

In this paper, we analyze the pre-processing methods used in automatic e-document classification of the above-mentioned types of e-documents, and elicit parameters that need to be taken into account for optimizing the pre-processing for each of these types of documents.

The remainder of this paper is organized as follows. Section 2 provides a review of the pre-processing phase in automatic e-document classification. Section 3 examines classification methods for e-catalogs, e-mails and web documents with emphasis on their respective pre-processing methods. Section 4 provides a comparative analysis of the pre-processing methods. Section 5 concludes the paper.

Pre-processing Parameters by e-Document Type

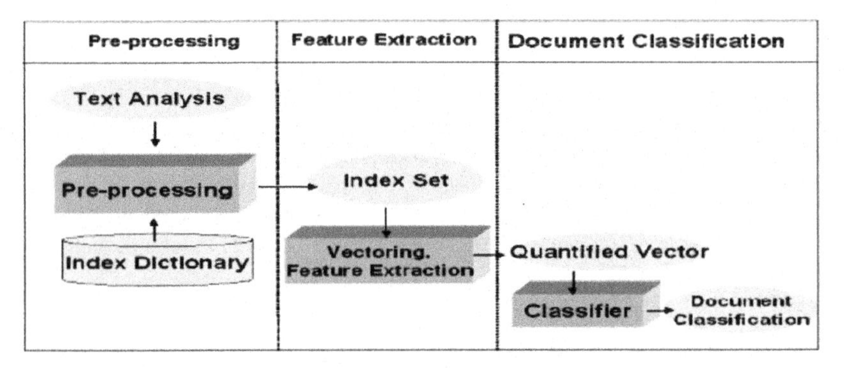

Fig. 1. The Three Phases in Automatic e-Document Classification

2 General e-Document Classification

The overall process of automatic e-document classification consists of three major phases, as shown in Figure 1. They include pre-processing, feature extraction, and document classification.

Many e-document classification methods require pre-processing as general text classification. In textual document classification, the first step in pre-processing is stemming through text analysis. Eliminating terms eliminates index words that appear with high frequency but that have little or no value in the meaning of the text. When eliminating index terms, we first need to create an index dictionary.

The next step is to convert multiple keywords with the same meaning in index set into a single keyword. A word consists of a stem that contains the meaning of the word and a prefix/suffix which changes the form of the words.

The aim of the vectoring step is to save a stem file instead of the words thereby reducing the volume of the index set file. The selection of index set for the feature extraction is very important to increase the accuracy of document classification.

The final phase is document classification. Classification is the process of assigning documents composed of quantified vector to previously defined categories by a classifier. In this phase, it is often necessary to observe users in order to generate rules for performing the actual classification. Previously established classification rules enable automatic document classification.

3 Pre-processing Methods for Three Types of e-Documents

3.1 e-Catalogs

E-catalog (Electronic catalog) hold information of products and services in an e-commerce system. E-catalog classification is the task of assigning an input catalog to one of the predefined categories (or classes). The categories are defined according to a specific classification scheme which usually has a hierarchical structure and thousands of leaf nodes. Correct classification is essential not only for data alignment and synchronization among business partners but also to keep the rapidly increasing product data maintainable and accessible. However, product data classification is a highly time-consuming task when it is done manually because of the increased scale of product information and the inherent complexity of the classification schemes. A number of competing standard classification schemes, such as UNSPSC and eCl@ss, do exist nowadays. Since none of them has yet been accepted universally, we need to re-classify the product data for mapping between different classification schemes as a part of business operations [2]. The reclassification task increases the need for an automated approach for product data classification.

E-catalogs are text intensive documents, so text classification techniques can be applied to e-catalog classification as well. Works in [3] applied several techniques from in-formation retrieval and machine learning to product data classification. Among the well known algorithms such as Vector Space Model, k-Nearest Neighbor, and Naive Bayes Classifier, they report that the Naive Bayes Classifier shows the best accuracy.

Although there is no single standard form of the e-catalog, conceptually it is a set of attribute-value pairs which, in turn, can be viewed as a structured document. In other words, instead of having product description comprised of one complete sentence, most e-catalogs are composed of an array of terms and numeric values aligned for each attribute, which has different discriminative power in classifying e-catalogs. Moreover, product information, unlike regular document, is frequently very short in length and composed of domain-specific terms, composite words, abbreviations and UOMs(unit of measure), which would be disregarded in regular document classifications.

Structural information, i.e. attributes, of e-catalogs are valuable information that should not be ignored during e-catalog classifications. E-catalogs of each class have a finite set of common attributes such as product name, supplier and model number, and a class-specific set of attributes such as height, length, diameter, color etc. Because of these characteristics, companies that build new catalogs face difficulties in building product classification and attribute systems. Recently, organizations publish the classification and attribute systems in terms of technical dictionaries such as eOTD, GDD, and RNTD of ECCMA, EAN/UCC, and RosettaNet, respectively [4][5][6]. Such class-specific structural information gives hints to a classifier, which is not available in regular document classifications.

As mentioned above, e-catalog has different set of features compared to regular document.

- Use of composite terms, abbreviations, and special terms: These special terms cannot be effectively used when the controlled vocabulary set, comprised of regular nouns, are defined as regular document's vector model.
- Short length of e-catalogs: We need semantic expansion of expressions in e-catalogs through using synonyms and lexical analysis of composite terms; otherwise, terms of trained e-catalogs would rarely match with those of new e-catalogs.
- UOM standardization: E-catalogs may be classified to other classes depending on product size, weight or other numeric attributes. It is critical to make full use of such attribute values comprised of numeric value and UOM., but the problem is that the UOMs used for product information are not standardized over the industries, and even the same UOMs can be expressed by different symbols; e.g. l, liter and _, and ohm and Ω.

Research for the use of the ontology for e-catalogs is active in the field of e-catalog to enable sharable, reusable and semantically enriched reference systems [7][8]. In the e-catalog ontology, various kinds of semantic information and relationships regarding product and product classifications are defined. Examples include relationships between products, attributes specific to each product class, products belonging to each class, UOMs for use by individual attributes, UOM conversion rules, lexical information such as standard term, abbreviation, synonyms, hyponyms, narrower terms and so forth. By leveraging the product ontology, it is possible to exchange product information with other e-catalog systems and integrate the e-catalogs.

This type of e-catalog ontology is leveraged to suggest a model that can execute automatic classification. However, it is not possible to systematically extract definite classification rules from them, since vocabulary and attribute do not definitely determine the classes. Accordingly, it is necessary to use a fuzzy or probabilistic approach, and the Naive Bayesian classifier is known to yield effective classification performance in classifying e-catalogs.

Pre-processing for e-catalogs starts with vocabulary standardization after stemming and the elimination of special symbols. Next, numeric values and UOMs are converted to standard UOMs by leveraging UOM conversion rules, defined in ontology. As for the generation of learning data, catalogs are extracted from all classification methods of Ontology (attribute, vocabulary) and subjected to analysis. Then, the frequency of appearance is calculated for vocabulary for each classification. On the other hand, the frequency of classification's appearance is calculated for vocabulary and attribute [7].

In the actual classification phase, e-catalogs contain attributes defined for individual classification. One catalog contains the value of an attribute, which in turn is comprised of a vocabulary aggregate (bag-of-vocabularies). Vocabulary is comprised of a term and the type of the term, and contains information such as whether the type is the original copy, a standard term that has been converted, whether it is a synonym or hyponym expansion, or a term that was subjected to parsing from a compound term, and so forth. Conversion based on UOM standardization assumes a type as a standard term, and the value of the original copy is eliminated.

Therefore, this research defined learning document by using MAD. In this measured value, uncertainty is measured by using the distance between above defined values of and these values $P(c\|x)$ average (μ). This is defined as follows.

Def.1. Product catalog P

$P = \{ (a, V) \mid a$ _ *Attribute* , V is a bag of vocabularies $\}$

Vocabulary V

$V = \{(v, t) \mid v$ *is a word and t* is the type of v$\}$

An existing Naive Bayesian classifier is expanded as follows to enable the assigning of a different weight by the type of an attribute and vocabulary in order to classify an e-catalog of a particular type. The hypothesis here is a Naive Bayesian assumption that all the attributes of the documents are independent.

Def .2. Naïve Bayesian Classifier

$$v_{NB} = \arg\max_{c_j \in C}\{P(c_j)\prod_{i,k} P((a_k, v_i) \mid c_j)\}$$

$$= \arg\max_{c_j \in C}\{P(c_j)\prod_{i} P(v_i \mid a_k, c_j)P(a_k \mid C_j)\}$$

In the above formula, $P(a_k\|c_j)$ expresses the weighted value of an attribute for a classification. To endow a vocabulary type with a weighted value, we estimate $P(a_k\|c_j)$ as follows.

n_{a_k,c_j} expresses the total number of value for an attribute keyword that applies to the total products within the learning data set, that corresponds to the attribute a_k of product category that is included in the class c_j. $Voca_k$ refers to the number of all values of attribute $a_k \cdot n_{(a_k,v_i)}$ shows the frequency of appearance for the value of attribute keyword vi that belongs to the class $c_j \cdot w_i^t$ shows the weight of type t of the vocabulary v_i. In conclusion, the probability of Def. 2 may be estimated as follows.

$$v_{NB} = \arg\max_{c_j \in C} \{ P(c_j) \prod_{i,k} P((a_k, v_i) \mid c_j) \}$$

$$= \arg\max_{c_j \in C} \{ \| c_j \| \prod_{i,k} \frac{n_{(a_k, v_i, c_j)} + 1}{n_{a_k, c_j} + | Voca_{a_k} |} \times w_i^t \times w_i^{a_k} \}$$

where $\|c_j\|$ denotes the size of the class c_j.

A diverse set of machine learning methods (k-Nearest Neighbor, Vector Space Model, Naive Bayesian Classifier) are used in [9] to compute product classification with the UNSPSC and their performance is compared. The Naive Bayesian Classifier was the winner. In [10], the question of integration of two catalogs is addressed, and applied the expanded Naive Bayesian Classifier.

3.2 E-mails

E-mails feature a far greater use of free style terms such as slang and abbreviations than any other types of documents. By identifying this particular characteristic of e-mails, and thus being able minimize the atypical aspects, the overall classification may be improved. Figure 2 shows the pre-processing phase for e-mail classification.

The followings are considered in the pre-processing phase of e-mails.

- Elements of potential error found in documents must be eliminated prior to indexing. The elements of error are those that obstruct the indexer. A representative example is a 2-byte symbol.
- Unnecessary information found in a document (a "noise" element) must be eliminated. Words that are generally used but have little meaning on their own must be controlled. This can be a key factor that obstructs judgment in document classification. A list of useless terms may be created and passed on to the indexer, so that more general index terms can be drawn out of the indexing process.
- In e-mail documents an HTML tag format is allowed. Accordingly, if the tag information is indexed together with e-mail's content, unnecessary index terms are produced, so the tags included in the e-mail should be eliminated.
- An e-mail document is made up of subject and main text. By leveraging this structural feature, classification performance can be improved.

To ensure precise automatic classification of e-mails, pre-processing is more important than for documents of other types, as it is necessary to minimize the non-standard features of e-mails,. In our prior research [11], we applied a 3-stage pre-processing algorithm.

In the first stage, documents for learning are selected in the feature extraction phase. An arbitrary aggregation of documents for learning is not used as is. Instead,

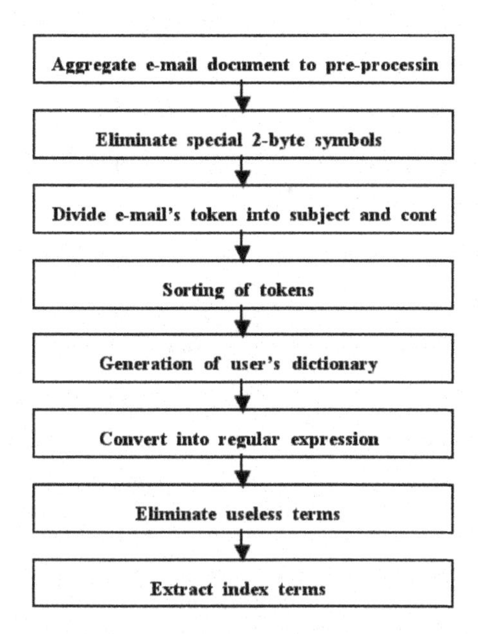

Fig. 2. The Pre-processing Phase in e-Mail Classification

Uncertainty-based Sampling Algorithm is applied, which in turn uses the MAD (mean absolute deviation) method that the computes the deviation of these values from their average. Uncertainty is measured by using the distance between values of $P(c||x)$ as conditional probability for each category c of one document x and their average (μ). This is defined as follows.

$$U_{MAD}(x) = \frac{1}{|C|}\sum_{i=1}^{|C|}(p(c_i \mid x) - \mu) \qquad \mu = \frac{1}{|C|}\sum_{i=1}^{|C|}P(c_i \mid x)$$

$U_{MAD}(x)$ refers to the average distance that showcases the deviation of these values from their average when it comes to the probability, subject to each category, or member values of document x. Thus, uncertainty is greater as $U_{MAD}(x)$ is lower, and vice versa. Here, average standard deviation of documents that are considered candidates for learning documents is calculated, and training data sets are selected with a priority set of documents with lower values.

In the second stage, features are extracted by assigning a weighted value for each attribute. E-mail document is comprised of title and main text. Weighted value assigning method by attribute is used to decide whether to assign weighted to weighted value on the title that stands multifold compared to the main text during the feature.

This method assigns a greater weighted value to the title than to the main text by applying an expanded Naive Bayesian Classifier.

$$P(v_i \mid c_j, a_k)P(a_k) \leftarrow \frac{n_{(a_k,l)} + 1}{n_{a_k} + \left| Voca_{sa_k} \right|} \times w_{sa_k}$$

n_{a_k} refers to the total number of keywords related to all attribute values that are applicable to the attribute a_k when it comes to the e-mail document that is included in the category c_j within the entire mail document of the learning data. $Voca_{sa_k}$ refers to the number of all keyword attribute values that are selected within the e-mail document in order to represent classification when it comes to attribute $a_k \cdot n(a_k, l)$ shows the frequency of appearance of the attribute value keyword v_i that belongs to the attribute a_k within the category c_j whereas w_{sa_k} manifests the individual weighted value for attribute a_k.

In the third stage, presumptive algorithm that conducts rule generation is cited as a decision factor for accuracy level of document classification. The role of this algorithm is to form ultimate rule by using learning document set, selected by configuration method for learning document set, and applies Naive Bayesian algorithm that uses dynamic threshold. A Naive Bayesian algorithm that uses a dynamic threshold value that improved on the existing fixed threshold value was applied.

3.3 Web Documents

Documents in the web are lack of logical organization. Besides, the enormous number of web documents make the manipulation and further operation on web documents difficult. Although the size of web document set is large, we need not analyze all the web documents as a whole. It can usually be divided into disjoint sets based on their document content. Web documents contains a set of hyperlinks that points to other web documents. These sets of hyperlinks can provide information about inter-relationship among web documents. We must use algorithms to partition a set of web documents based n their networked hyperlink structure [12]. Also, Web documents mostly use HTML.

Anyway, most of the classification methods for Web documents are comprised of three steps. First is monitoring of user's behavior. Second is the creation of a user profile based on the monitoring results. Third is providing Web documents that the user is interested in by leveraging the user profile.

Pre-processing for Web document has the HTML tag elimination step in addition to the pre-processing steps for general documents and the assignment of weighted value according to hyperlink's level

- HTML tag elimination: Web documents are created in the HTML format. As such, the processing of the HTML tags is necessary. An HTML tag is surrounded by '<', '>', and the elimination of the tags is necessary for the selection of precise keywords in Web documents.

- Procedure afterwards requires the elimination and stemming of unnecessary terms as for general documents.
- Feature extraction: Feature extraction assigns a weighted value for each feature extracted according to the level of Hyperlink.

We have conducted research on the Web document classification system, and a representative system case includes Personal Webwatcher, InfoFinder [13] and NewT [14] etc.

4 Comparative Analysis of Pre-Processing Methods

Thus far, we have examined the pre-processing methods that take into consideration e-document characteristics. Table.1 summarizes the results of our discussions. There are many parameters in pre-processing; the removal of HTML tags, the removal of special signs that play the role of stop words, the processing of unnecessary words, the streaming processing that factors in changes in meanings even when the stem is the same, the standardization of vocabularies, the assignment of weighted values, and additional parameters that are specific to each type of document.

	Pre-processing Parameter						
Document Type	Elimination of Tag	Elimination of Special Symbols	Processing of Unnecessary Terms	Stemming Processing	Vocabulary Standardization	Assigning Weighted Value	Additional Processing
General Document	-	-	O	O	O	-	-
Web Document	O	-	O	O	O	-	Calculation of Hyperlink's level
E-mail Document	-	O	O	O	O	O	-
E-Catalog Document	-	X	-	-	O	O	UOM Processing, Expansion of Information

Table 1. Pre-processing Parameters by e-Document Type

When a document set is subjected to automatic classification, it is necessary to check all the necessary pre-processing parameters, and to sufficiently factor in these parameters as summarized in Table 1. As such, it is possible to increase performance when it comes to classification by executing the pre-processing method for a particular document type.

In section 3, we laid out each of the different pre-processing phase for document classification. As a result, we were able to show the crucial pre-processing parameters into table 1. As illustration in table. 1, the necessary pre-processing parameters for each of the documents are different.

In the case of Web documents, HTML tag elimination and Hyperlink's level calculation is needed. E-mail documents are less formal and thus requires a more complicated pre-processing comprised of many phases. Here, the phase of assigning weighted value to the title of the e-mail documents is especially important. Finally, for e-Catalog Documents which are composed of short terms related to product and product units, pre-processing is comparatively simple yet it requires additional processing.

This can be illustrated by the next example. Using the pre-processing parameters in table.1, we will show how the different document characteristics are applied to e-Catalog document.

Pre-processing Parameters by e-Document Type

```
<catalog id="36275910 ">
        <product-name> IBM ThinkPad G41 </product-name>
        <desccription> Intel P4 3.0GHz, ATI X300, HDD 60G </description>
        <unspsc-code>43171802</unspsc-code>
        <unspsc-name>Notebook computers</unspsc-name>
        <producer>IBM</producer>
        <individual-attr>
                <attribute id="CPU"> Intel Pentium4 3.0GHz </attribute>
                <attribute id="HDD"> 60Gigabyte </attribute>
                <attribute id="graphics"> ATI X300 </attribute>
                <attribute id="memory"> 512M </attribute>
                <attribute id="weight"> 2.5kg </attribute>
                <attribute id="display size"> 14" </attribute>
        </individual-attr>
</catalog>
```

Fig. 3. An example of e-catalog

Figure 3 shows an example of e-catalog in XML format. As the example shows, attributes of products appear as tags in the e-catalogs. Although tags in HTML documents, e.g. <title>, <head> and <table>, does not imply any semantics, those of e-catalogs representing attributes carry class-specific information, thus should not be eliminated during pre-processing. Furthermore, each attribute has a different discriminating power and it can be weighted differently during classification. Special symbols also carry valuable information in e-catalogs. Consider the display size of the notebook computer in Figure 3 represented by 14". The double-quotation marks, usually eliminated in text classification, indicate that the product has an attribute denoting a length which is obviously a good hint for classifying the product.

Stemming and processing of unnecessary terms, however, have little application since the values in e-catalogs are short and itemized noun phrases rather than long descriptive sentences. Vocabulary standardization is useful in e-catalog classification as is in other domains. For example, if notebook computers, laptops and portable computers are standardized to laptop computers, it is obvious that the classification algorithm will be improved. Finally, UOMs are frequently used in describing product detail. If the classification algorithm understand that 2.5kg is same as 5.5 lb, the performance gain in e-catalogs would be greater than in text classification since the product classification are generally more dependent on the numeric values.

Thus, as seen from the example of e-Catalog documents, table 1 can be used to determine which phases are necessary and which are not when classifying a certain type document.

Although the parameters in the table may seem simple and even similar to each other, it makes it possible to know the pre-processing phases in advance so that accurate and efficient classification can be achieved.

5 conclusion

We have elicited seven key parameters that play major roles in the performance of the pre-processing phase of automatic e-document classification for three major types of document, namely, e-catalogs, e-mails, and Web documents. The key parameters were derived by taking into consideration the particular characteristics of the e-documents. The results of our research, which is summarized in Table 1, can serve as a guide for selecting key parameters for implementation and optimization for effective automatic classification of the three most important types of e-document in use today. Our Proposal is to improve the pre-processing phase in automatic e-document classification by considering the document-specific characteristics, while continuing to exploit the state of the art classification algorithms. We expect that the pre-processing techniques can be applied to other e-document types that exhibit similar characteristics.

Acknowledgement. This work was supported by the Ministry of Information & Communications, Korea, under the Information Technology Research Center(ITRC) Support Program.

References

1. Chidanand Apt, Fred Damerau, and Sholom M. Weis, "Towards Language Independent Automated Learning of Text Categorization models," proc. of the 17th annual international ACM-SIGIR, 1994.
2. D. Fensel, Y. Ding, E. Schulten, B. Omelayenko, G. Botquin, M. Brown, and A. Flett: Product Data Integration in B2B E-commerce. IEEE Intelligent System, Vol 16/3, pp 54-59, 2001.

3. Y. Ding, M. Korotkiy, B. omelayenko, V. Kartseva, V. Zykov, M. Klein, E. Schulten, D. Fensel: GoldenBullet: Automated Classification of Product Data in E-commerce. Business Information System 2002, 2002.
4. eOTD, ECCMA Open Technical Dictionary, ECCMA, http://www.eccma.org/eotd/-index.html.
5. GDD, Global Data Dictionary, http://www.ean-ucc.org/global_smp/global_data_ dictionary.htm.
6. RosettaNet, The RosettaNet Technical Dictionary, http://www.resettanet.org
7. Ig-hoon Lee, Suekyung Lee, Taehee Lee, Sang-goo Lee, Dongkyu Kim, Jonghoon Chun, Hyunja Lee, Junho Shim, "Practical Issues for Building a Product Ontology System", DEEC 2005 (Data Engineering Issues in E-Commerce).
8. L. Obrst, R.E. Wray, and H. Liu, "Ontological Engineering for B2B E- Commerce", International Conference on Formal Ontology in Information Systems (FOIS'01), ACM Press, 2001.
9. Y. Ding, M. Korotkiy, B. Omelayenko, V. Kartseva, V. Zykov, M. Klein, E. Schulten, and D. Fensel, "GoldenBullet: Automated Classification of Product Data in E-commerce," Proceedings of Business Information System, 2002
10. R. Agrawal and R. Srikant, "On Integrating Catalogs", the 10th International World Wide Web Conference, 2001
11. Ok-Ran Jeong, Dong-Sub Cho, "A Rule Filtering Component based on Recommendation Agent System for Classifying Email Document", Volume LNCS 3320, Springer Verlag, pp. 697-703, Singapore, December 2004
12. A.H. Tan and P. Yu(Eds): PRICAI 2000 workshop on Text and Web Mining, Melbourne, pp.44-51, August 2000
13. Haejung Bak, Yeongdaek Park, Sukhwan Yun, "Web Agent using user's favorite", Korea Information Processing Society Review, September 1999.
14. H. Nwana and N. Azami, "Software Agent and Soft Computing: Towards Enhancing Machine Intelligence," Springer, pp. 160-179, 1997

Packaging Web Product Information in XML for Mobile Clients

Jung-Su Hur and Sangho Ha

School of Computer Science and Engineering,
Soonchunhyang University,
Asan-si, Choongnam, 336-745, Korea.
{jshur,hsh}@sch.ac.kr

Summary. Advances in wireless technologies and mobile computing are enabling many kinds of mobile devices to be used for m-commerce. Service content for m-commerce is usually newly written to meet specific characteristics of the target mobile devices. To avoid these formidable efforts, it is very important to effectively exploit the Internet merchant information currently served by e-commerce. However, bringing the Internet content to mobile devices is far from straightforward due to the many limitations of mobile devices such as little memory, small displays, and low processing speeds. In this paper, we suggest a system to extract only the product elements interesting to users from HTML documents of the Internet shopping malls, package them in XML, and deliver the result to mobile clients. We then apply our system to the sample HTML documents.

1 Introduction

M-commerce [1] is being realized using several kinds of mobile devices such as PDAs, mobile phones, and Smart phones. Examples of m-commerce today are NTT DoCoMo's i-Mode portal [2], Nordea's WAP Solo Mobile Banking Service [3], and Webraska's SmartZone Platform [4]. Service sites for m-commerce are usually designed to meet specific characteristics of the target mobile devices as with i-Mode. For example, i-Mode has over 1, 000 official content providers. They develop their own i-Mode sites where contents are written specifically for their mobile services using Compact HTML(cHTML) [5], which is a subset of HTML. These contents can not be directly served on mobile devices other than i-Mode. If they are to be rewritten to meet limitations and idiosyncrasies of another device, the required modifications are formidable and redundant.

In addition, the Internet services could not be directly served for mobile services. Several studies [6][7][8] have been done to deliver Web content to mobile devices by applying several techniques such as scaling, manual authoring, transducing, and transforming. Related work can be also found in the personalization of web pages for mobile clients such as WebViews [9], PWA [10], and Wiccap [11]. In these studies, Web users create their own customized views of Web sites that are well-suited for

J.-S. Hur and S. Ha: *Packaging Web Product Information in XML for Mobile Clients*, Studies in Computational Intelligence (SCI) **110**, 27–35 (2008)
www.springerlink.com

their own mobile devices. Those Web views are thus considered only for an individual mobile client. This research did not consider exploiting the Internet product information for m-commerce. Since the product information is structured differently than typical Web content, we need to tailor the information to provide mobile clients with user-friendly views.

Previously, we developed a system [12] to effectively exploit the Internet product services for clients with mobile phones. We assumed in the work that the Internet product information was written in XML. This assumption, however, is not realistic because many shopping malls are still written in HTML.

In this paper, we suggest a system to extract the product information written in HTML from the Internet shopping malls and deliver it to mobile clients. The extracted product information is packaged in XML before it is delivered to mobile clients. The system has three main parts: the Internet shopping malls, server, and mobile clients. Server manager first uses the specially designed GUI(Graphical User Interface) to mark product elements essential to mobile clients from HTML documents of the Internet shopping malls. The GUI gives the manager the pop-up menu showing product elements to be marked. We use the simplified but general product structure [12] designed for mobile clients based on electronic catalogue [13], which is called the 3MP model. The manager finds and collects product elements of the structure from HTML documents. The path information to those product elements from the top element of HTML documents is calculated and stored in the local database. On user request, the path information is used to extract the marked product elements from HTML documents of the Internet shopping malls. The extracted product elements are collected into the 3MP model, packaged in XML, and delivered to mobile clients. In section 2, we will design a system to extract the product information from HTML documents and package it in XML. In section 3, we will apply our system to the sample HTML documents. Finally, we will summarize our work in section 4.

2 System

Figure 1 shows the server architecture of our system. The server has a role of the interface between the two parts: the Internet shopping malls and mobile clients. We want the server to first extract the product information from HTML documents of the Internet shopping malls requested by mobile clients, package it in XML, and send the packaged information to the mobile clients. The Internet shopping mall documents are usually written in HTML. These HTML documents are packaged in XML by the two steps. The first step is a marking phase. In this phase, product elements in which mobile clients are essentially interested are marked in the HTML documents. The second step is an extraction phase. In this phase, the marked product elements are extracted from the corresponding HTML documents and are packaged in XML. To extract the marked product elements, we maintain the path information of the marked product elements in HTML documents. We also use the product structure defined in [12], called the 3MP model, to organize product elements effectively.

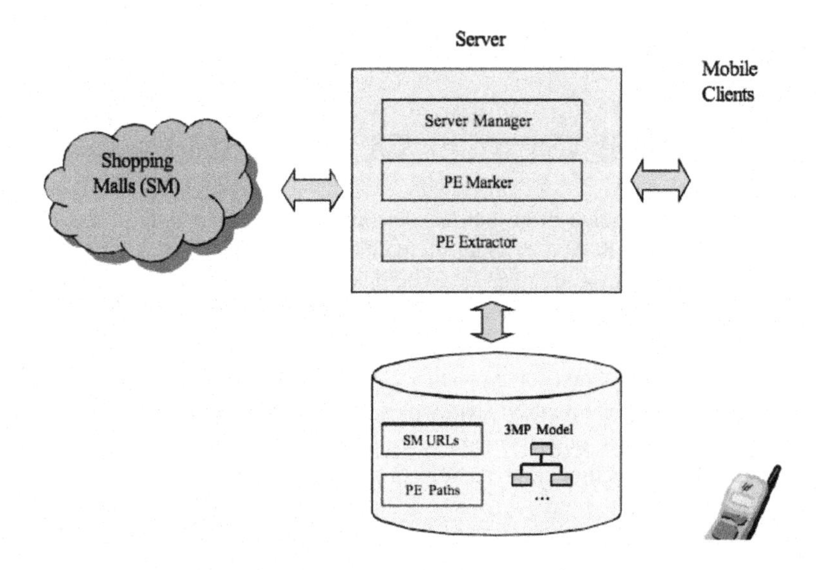

Fig. 1. The Overall System Architecture

As shown in Figure 1, the server includes three components: Server Manager, PE Marker, and PE Extractor. PE Marker has a role of marking product elements in the HTML documents in which mobile clients are mainly interested, and PE Extractor has a role of extracting the marked product elements from the corresponding HTML documents and packaging them in XML. We will below describe each component in detail.

2.1 Server Manager

It registers the Internet shopping malls and processes user's query. The registration of the Internet shopping malls is done simply by storing their URL in the database. The server manager gets search words from user's query and the relative shopping mall URLs from the database, and combines them to generate search queries. Each search query is sent to the corresponding shopping mall site. User queries can include advanced query conditions such as the range of prices and the preferred product makers for the personalization, which can be done by further filtering the product information by the user's conditions given by the advanced query.

2.2 The 3MP Model

It is an electronic catalog [12] designed specially for m-commerce. It is actually a compact version of the electronic catalog [13] designed for e-commerce, including only essential elements that mobile clients are mainly interested in. However, it is so general that it can be used for describing all kinds of products in the Internet shopping malls. This model is referred by PE Marker and PE Extractor in our system. The XML documents packaged from HTML documents thus conform to the 3MP model.

2.3 PE Marker

It marks product elements to be extracted from HTML documents of the Internet shopping malls. The marked product elements are indicated in the HTML source code. A brief description about how to mark product elements is given below. A specially devised GUI is used for these marks. A system administrator first loads a sample HTML document of a shopping mall on a standard browser such as Internet Explorer. When he finds some product element to be extracted in the document, he selects the product element by shadowing it in the document and clicks the right button of his mouse. A pop-up menu then comes up on the click position. It exactly reflects the 3MP model. If he selects an item in the menu that matches the product element selected already in the document, a specific tag is prefixed to the selected product element in the HTML source code of the document. We designed tags to be prefixed in the HTML source code. They are considered for elements in the 3MP model. For example, for the maker element in the 3MP model, two tags are considered: xml_maker and xml_alias_maker. The maker element in the model is used to represent makers of products. When markers of some products are marked in a HTML document, xml_marker is prefixed to the marker content in the HTML source code. An xml_alias_marker can be prefixed to the synonym which denotes "maker of products". Notice that words denoting makers of products can be various, depending on types of products. For example, a word of publisher is used to denote makers of books while a word of manufacturer is used to denote makers of computers.

Since HTML documents of shopping malls usually have the same document structure for the same product category, we can group HTML documents of a shopping mall by the product category. Each group then includes HTML documents with the same structure. Marking product elements in HTML documents is done for each group of each shopping mall registered in the database.

2.4 PE Path Calculator

It is a part of PE Marker, and calculates paths of product elements in HTML documents which are marked by PE Marker and stores them in the local database. The path is represented by a sequence of HTML tags from the initial tag of HTML documents to the marked product element. It is also associated with the special XML tag prefixed to the product element. As a result, a pair of an XML tag and a path is stored for each product element marked in a HTML document. We now can use the path to extract the marked product element from a HTML document, and use the XML tag to interpret the meaning of the extracted product element. We call the pair the PE locator. Notice that PE locators for the marked product elements can be indexed by a pair of a shopping mall URL and a category.

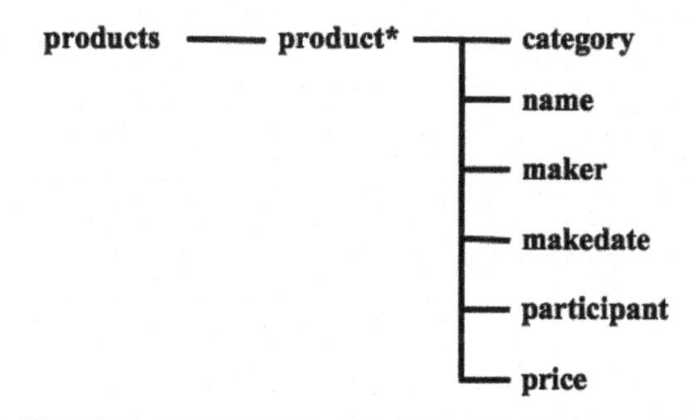

Fig. 2. The Simplified 3MP Model

2.5 PE Extractor

HTML documents of the Internet shopping malls usually include a lot of information which is unnecessary to mobile clients. PE Extractor extracts only product elements from HTML documents which is essential to mobile clients, and arranges them into an XML document conforming to the 3MP model. Notify that product elements essential to mobile clients was already marked by PE marker. For the incoming HTML documents, it first gets a set of PE locators of the group of the HTML documents from the database. It then gets a PE locator from the set in sequence, and uses it to extract the corresponding product element from the HTML document. The meaning of the extracted product elements can be interpreted from their associated XML tags. We can arrange the extracted product elements by the 3MP model, based on their meaning. As the result, we get an XML document conforming to the 3MP model, including only the product elements extracted from HTML documents. XML documents generated from HTML documents with the same category can be combined into a single XML document.

3 Application

We here apply our system over sample HTML documents in the simplified form. Figure 2 shows the simplified version of the 3MP model [12]. class denotes a class of products according to a specific classification scheme, name denotes the name of products, maker and makedate denote the manufacturers and the manufactured date of product, respectively, participant denotes persons who directly participate in manufacturing the product, and price denotes the price of products. Figure 3 shows an example HTML document and product elements to be extracted from the document.

We will now show how to mark product elements to be extracted from the example HTML document, extract them from the document, and package them in XML in sequence. Figure 4 shows how to mark product elements in HTML documents using the GUI devised in this paper. In the figure, user shadowed a product element

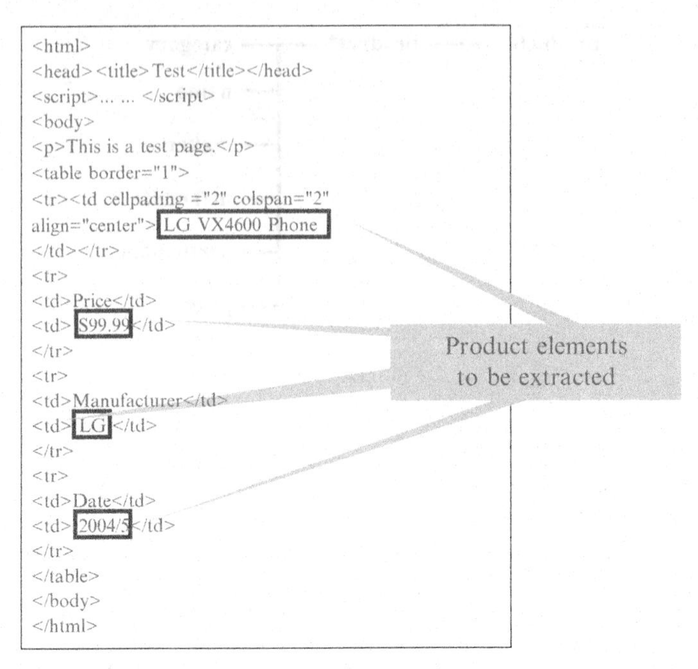

```
<html>
<head><title>Test</title></head>
<script>... ... </script>
<body>
<p>This is a test page.</p>
<table border="1">
<tr><td cellpading ="2" colspan="2"
align="center">LG VX4600 Phone
</td></tr>
<tr>
<td>Price</td>
<td>$99.99</td>
</tr>
<tr>
<td>Manufacturer</td>
<td>LG</td>
</tr>
<tr>
<td>Date</td>
<td>2004/5</td>
</tr>
</table>
</body>
</html>
```


Fig. 3. An Example HTML Document

of "LG", and clicked the right button of his mouse using the GUI. We can see a pop-up menu come up in the clicked position. The menu shows the tree structure of the 3MP model exactly. When he selects two items of marker and value in sequence, the selected product element is marked accordingly. Figure 5 shows the resulting HTML code after marking all of four product element denoted to be extracted in Figure 3. We can see three xml-alias tags prefixed additionally in the document, which will provide users with the more precise meaning and representation for the corresponding product elements.

Figure 6 shows the set of PE locators calculated for the marked HTML document in Figure 5. We consider the set of PE locators for the product element of name. It is represented by an array. An index of the array except for the first element represents the depth in the HTML tree structure. For example, the element with the index value of 1 represents that the html tag is located the first level of the tree. The element is also associated with the order of siblings in the same depth level of the tree. That is, html(1) means that it is the first sibling in the first depth level. Similarly, body(3), which is the element with the index value of 2, means that it is the third sibling in the second depth level of the tree. Therefore, we can see that the element associated with the xml_name tag is located within the third child tag, body, of the html tag, within the second child tag, table, of the body tag, within the first child, tr, of the table tag, and within the first child, td, of the tr tag in the HTML code. The associated xml_name tag gives us the meaning of the corresponding product element. Figure 7 shows the resulting XML document generated by PE Extractor for HTML documents coming from the Internet shopping malls. We can see that PE Extractor

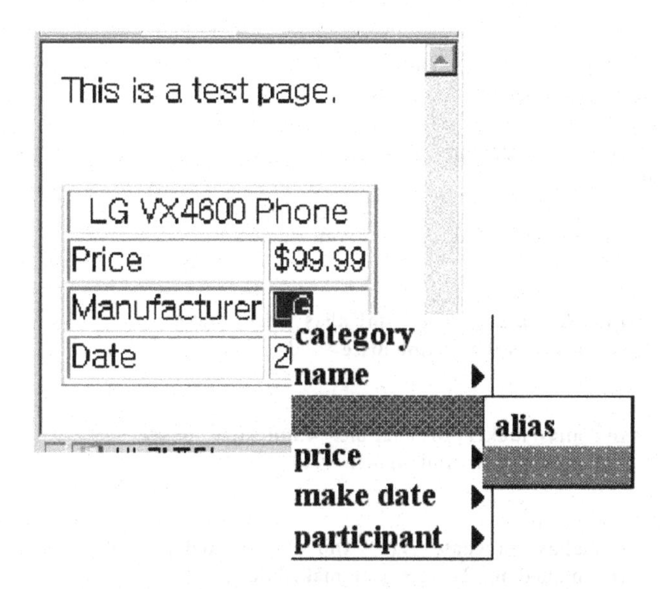

Fig. 4. A GUI for Marking PE

extracted product elements from three HTML documents with the same structure as in Figure 3 using PE locators in Figure 6, and packaged them into an XML document conforming to the 3MP model.

4 Conclusions

In this paper, we suggested a system to extract only the product information concerned with users from HTML documents of the Internet shopping malls, and deliver it effectively to mobile clients. Since the extracted information is represented in XML, it can be used in any kinds of mobile devices. If mobile devices can not handle XML documents, the information can be transformed into the readable form by the devices before it is delivered. The system can be used under minor changes of HTML documents as long as paths of the marked product elements are not changed by the modification of HTML documents. Major changes of HTML documents can be detected by comparing the extracted element with the marked element. In that case, the set of path locators of the corresponding HTML documents should be calculated again. The system is being implemented on the J2EE platform [14]. We have a plan to integrate the product information across multiple shopping malls to provide clients with more added information.

Acknowledgement. This work was supported by grant No. R05-2004-000-12565-0 from Korea Science & Engineering Foundation.

```
<html>
<head> <title> Test</title></head>
<script>... ... </script>
<body>
<p>This is a test page.</p>
<table border="1">
<tr><td cellpading ="2" colspan="2" align="center">
<xml_name> LG VX4600 Phone </xml_name>
</td></tr>
<tr>
<td><xml_alias_price>Price</xml_alias_price></td>
<td><xml_price>$99.99 </xml_price></td>
</tr>
<tr>
<td><xml_alias_maker>Manufacturer </xml_alias_maker></td>
<td><xml_maker>LG</xml_maker></td>
</tr>
<tr>
<td><xml_alias_makedate>Date</xml_alias_makedate></td>
<td><xml_makedate>2004/5 </xml_makedate></td>
</tr>
</table>
</body>
</html>
```

Fig. 5. The Marked HTML Code

xml_name	html(1)	body(3)	table(2)	tr(1)	td(1)
xml_alias_price	html(1)	body(3)	table(2)	tr(2)	td(1)
xml_price	html(1)	body(3)	table(2)	tr(2)	td(2)
xml_alias_maker	html(1)	body(3)	table(2)	tr(3)	td(1)
xml_maker	html(1)	body(3)	table(2)	tr(3)	td(2)
xml_alias_makedate	html(1)	body(3)	table(2)	tr(4)	td(1)
xml_makedate	html(1)	body(3)	table(2)	tr(4)	td(2)

Fig. 6. A Set of PE Locators

References

1. Norman Sadeh, M-Commerce: Technologies, Services, and Business Models, Reading, Wiley, 2002.
2. http://www.nttdocomo.com/.
3. http://www.nordea.com/.
4. http://www.webraska.com/.
5. Deitel, Wireless Internet & Mobile Business-How to Program, Reading, Prentice Hall, 2002.

```
<?xml version="1.0"?>
<products>
        <product>
                <category/>
                <name>LG VX4600 Phone</name>
                <maker alias="Manufacturer">LG</maker>
                <price alias="price" location="US">99.99</price>
                <makedate>2004/5</makedate>
                <participant/>
        </product>
        <product>
                <category/>
                <name>Samsung E715 Phone</name>
                <maker alias="Manufacturer">Samsung</maker>
                <price alias="price" location="US">224.99</price>
                <makedate>2004/3</makedate>
                <participant/>
        </product>
        <product>
                <category/>
                <name>Motorola i730 Phone</name>
                <maker alias="Manufacturer">Motorola</maker>
                <price alias="price" location="US">124.99</price>
                <makedate>2004/1</makedate>
                <participant/>
        </product>
</products>
```

Fig. 7. The Packaged XML Document

6. A. Fox et al., Experience with top Gun Wingman, A Prox-Based Graphical Web Browser for the 3Com PalmPilot, Proc. IFI International Conf. Distributed Systems Platforms and Open Distributed Processing(Middleware 98), N.Daview, K. Rymond, and J. Seitz, eds. Springer-Verlag, London, 1998, pp. 407-424.
7. T.W. Bickmore and B.N. Schilit, Digestor: Device-Independent Acces to the World Wide Web, Computer Networks and ISDN Systems, vol. 29, nos. 8-13, 1997, pp. 1075-1082
8. O. Buyukkokten et al., Power Browser: Efficient Web Browsing for PDAs, Proc. Conf. Human Factors in Computing Systems(CHI 00), ACM Press, New York, 2000, pp. 430-437
9. Juliana Freire, et al., WebViews: Accessing Personalized Web Content and Services, In Proc. of WWW10 in LNCS, 2001.
10. L.T. Passos and M.T. Valente, Personalizing Web Sites for Mobile Devices Using a Graphical User Interface, In Proc. of ICWE in LNCS, 2004.
11. Zehua Liu, et al., Personalized Web views for Multilingual Web Sources, IEEE Inter-net Computing, 2004 July.
12. Sangho Ha and In-Gook Chun, Adaptation of the Internet Product Information for Mobile Clients, In Proc. of International Conf. on Internet Computing, 2004.
13. Sangho Ha and Kunsu Suh, A XML Based Product Description Model, International Conf. On Software Engineering, Artificial Intelligence, Networking & Parallel/Distributed Computing, 2001.
14. Rick Cattell, and et al., J2EE Technology in Practice, Reading, Addison Wesley, 2001.

Designing an Auction Protocol for Cases Involving Insincere Sellers: Motivating buyers to investigate unidentified sellers

Shigeo Matsubara and Masafumi Matsuda

NTT Communication Science Laboratories, NTT Corporation
2-4 Hikaridai, Seika-cho, Soraku-gun, Kyoto 619-0237, JAPAN
{matsubara, masafumi}@cslab.kecl.ntt.co.jp,
WWW homepage:
http://www.cslab.kecl.ntt.co.jp/csl/sirg/people/matubara/

Summary. This paper focuses on single-item multi-unit auctions involving multiple sellers, particularly in cases where sellers might be insincere, i.e., might defect in the middle of the process of exchanging goods and money. To exclude insincere sellers and attain efficient allocation, a buyer should investigate unidentified sellers and share information about them. However, the first buyer may suffer a loss if the seller is insincere, which discourages a buyer from acting as the first buyer. In addition, an auctioneer has to induce buyers to reveal true information about sellers as well as the true valuation of goods. To solve this problem, we have developed an auction protocol that motivates buyers to investigate unidentified sellers by determining an appropriate payment amount and schedule. In addition, the protocol recalculates the allocation of goods whenever new seller information is obtained. We prove that the protocol achieves equilibrium for buyers' truth-telling regarding information on both sellers and the goods.

1 Introduction

Trade in network environments includes various kinds of uncertainty. In designing a trading mechanism, a mechanism-design approach shows much promise [1], and AI can provide solutions to the problem of effectively dealing with uncertainty. For example, Yokoo *et al.* discussed a problem where the uncertainty of a participant's identity made an auction vulnerable [2]. Ito *et al.* studied cases where product quality is uncertain and proposed an auction protocol that attains efficient allocation by eliciting expert knowledge on the product's quality [3, 4]. Porter *et al.* discussed a mechanism design for cases where task execution might fail [5]. Matsubara studied auctions in a dynamic environment where the valuation of the good depends on environmental conditions [6].

This paper focuses on seller uncertainty, especially the case of a seller's intentional failure, i.e., fraud. On the other hand, Porter *et al.* dealt with exogenous failure

S. Matsubara and M. Matsuda: *Designing an Auction Protocol for Cases Involving Insincere Sellers: Motivating buyers to investigate unidentified sellers*, Studies in Computational Intelligence (SCI) **110**, 37–59 (2008)
www.springerlink.com

[5]. Online deals have become risky, since there exists much fraud in the process of exchanging goods and money. For example, there have been reports of auction winners sending large sums of money to sellers who then disappear without sending any goods [7]. Similar risks may exist among computational agents. Escrow services are available in the real world market, and solutions have been proposed that utilize computational agents for such tasks [8, 9, 10]. To eliminate the risk of fraud, however, a single technique is probably not sufficient, and thus having multiple forms of protection is desirable.

One way of excluding frauds is to use a reputation-based system [11, 12, 13]. In auction sites, multiple units of the same good are sold by different sellers, and the sellers' reputations are used by buyers to select sincere sellers. However, reputation has been discussed apart from the challenge of developing a good allocation process itself [14]. This paper tries to find a way of allocating goods as well as a way of eliciting the seller's reputation from buyers. Another related work can be found in [15], but our work differs from that study in that we deal with multiple sellers and fraud in the process of exchanging goods and money.

This paper deals with single-item multi-unit auctions involving multiple sellers and buyers. Our mechanism-design problems include the following: (1) Information on sellers cannot be obtained unless a buyer commits himself/herself to an actual deal; and (2) Two sources of information, i.e., valuation of the goods and seller information, must be dealt with simultaneously.

The former issue arises in network environments because it is difficult to directly observe remote sellers and learn their identity. Thus, buyers cannot gauge the sincerity of sellers unless they carry out a trade. Therefore, the first buyer to purchase from a particular seller faces a riskier situation than subsequent buyers because information on that seller is unavailable before the sale; consequently, no buyer is motivated to be the first buyer from that seller.

The latter issue means that attaining efficient allocation requires eliciting seller information from corresponding buyers as well as ascertaining the buyers' valuations of goods. A simple method for allocating goods fixes allocation at the beginning of trading and then not changing it, even if new seller information is obtained. In this case, buyers tell the truth about the seller's reputation because lying does not benefit buyers. However, buyers might be motivated to lie about their valuations of goods if the allocation dynamically changes. In addition, fixed allocation results in inefficient allocation; a buyer with a higher valuation sometimes cannot win the good while another buyer with a lower valuation can.

To solve this problem, this paper proposes a novel auction protocol with the following features: (1) it can motivate buyers to investigate unidentified sellers by setting an appropriate payment amount; (2) it can induce buyers to tell the truth about sellers' reputations (good or bad) by dynamically reallocating goods; and (3) it can attain an efficient allocation.

Section 2 describes the outline of mechanism design. Section 3 describes the model of an auction involving insincere sellers. In section 4, we propose a new protocol that can attain efficient allocation. Section 5 describes theoretical analysis, Section 6 discusses a seller's strategic behavior observed on eBay, and Section 7 concludes the paper.

2 Outline of Mechanism Design

In this section, we briefly explain what mechanism design is. Readers who want to know precise definitions of mechanism design should refer to [16]. Mechanism design means the approach of implementing a social choice function in the following situation. Multiple agents and a center exist. The center collects messages from the agents. Based on the collected messages, the center makes a social decision, which affects the agents' utilities. Here, agents are assumed to be self-interested. The problem is whether the social choice function that is desirable for the system designer can be implemented, i.e., whether the system designer can elucidate the agents' private information and thus make a decision that brings a desirable outcome.

In the case of auctions, bidders play the roles of agents and an auctioneer or a seller plays the role of the center. The auctioneer collects bids from bidders and then uses them to decide the allocation of the goods and the bidders' payments. In the case of voting, voters play the roles of agents and an election administration committee plays the role of the center. The election administration committee collects the voting ballots and uses them to decide the election results. Here, in the former case, there are various ways of deciding the goods' allocation and the payments, and different mechanisms have different properties in terms of allocation efficiency, seller's revenue, etc. In the latter case, there are various ways of deciding the election results, and different mechanisms have different properties in terms of anonymity, fairness, etc.

The research on mechanism design originated in economics, but it also contributes to design mechanisms in information systems. A P2P file sharing system involves a free-rider problem [17], usage of network bandwidth involves a tragedy of commons [18], and routing in a multi-hop network involves a hidden-action problem [19]. That is, incentive problems can be observed in many situations.

Compared to ordinary optimization problems, a technical difficulty in mechanism design is that agents are self-interested, i.e., the center has to elucidate agents' private information such as their valuations of the goods. If an agent is self-interested, he/she may manipulate the allocation and/or an amount of payment by misrepresenting his/her valuation to increase his/her utility. The center has to prevent agents' taking strategic behaviors to obtain a better outcome. This is an essential difficulty in designing a desirable mechanism.

In the rest of this section we focus on auctions and outline the factors that should be examined in designing a mechanism: preferences, objective function, constraints, and equilibrium. Readers who want to know auction theory in detail should refer to [20, 21, 22].

2.1 Preferences

In designing a mechanism, we first have to define bidders' preferences. In the context of auctions, preferences correspond to the valuations of the goods. We start discussion about value models. Value models are classified in the following three models [23].

- private value model
- common value model
- interdependent value model

A private value model represents the case where a bidder's valuation of goods is independent of those of the the other bidders. For example, we can assume that bidders' valuations of an antique are represented by using a private value model if the bidder's objective is not to resell it.

A common value model represents the case where the valuations of all the bidders are the same as those of each other but each bidder does not know the accurate value. For example, bidders' valuations of an oil mining right can be represented by using a common value model. This means that the amount of oil production is the same regardless of who excavates the land over the oil but different bidders have different estimates of this amount, e.g., a bidder who has excavated the area next to the auctioned area is likely to have a more accurate estimate than other bidders.

The private value model and the common value model are two extreme cases. However, the two models include different metrics, i.e., whether bidders' valuations are the same as one another and whether the bidder's valuation is accurate. Many real-world cases might be suitable for modeling as an interdependent value model in which different bidders have different valuations and a bidder's valuation is affected by the other bidders' valuations. However, analyzing an interdependent value model is often difficult. Therefore, many previous works have discussed a private value model or a common value model as a kind of approximation of the real world.

2.2 Objective function

In auctions, an objective function is used to maximize the social surplus or maximize the seller's revenue. The social surplus means that the sum of the seller's and all bidders' utilities is maximized. In the definition of utilities, we often assume a quasi-linear utility in which a bidder's utility is defined as the difference between the valuation of the allocated goods and the payment. When the number of the good to be sold is one, the social surplus is maximized if the good is awarded to the bidder having the highest valuation of the good.

Satisfying the condition of maximizing the social surplus is also called Pareto efficiency, which means a bidder's utility cannot be increased without reducing the seller's or some other bidders' utilities. In an auction, if Pareto efficiency is satisfied, social surplus is maximized because sellers and bidders can complete a money transfer. Here, we assume quasi-linear utility.

If the seller is the government, e.g., engaged in the disposal of government property, the government may choose to maximize the social surplus instead of maximizing its revenue. In addition, in complicated cases such as combinatorial auctions, the objective function is often set to maximize the social surplus because analyzing the revenue maximization is not tractable.

2.3 Constraint

In maximizing the objective function, the following constraints will be examined.

- individual rationality
- incentive compatibility
- budget balance

Individual rationality is indispensable, while violation of budget balance might be allowed. Incentive compatibility is useful in reducing the search space.

Individual rationality means that a bidder does not suffer any loss by participating in an auction as long as he or she is rational. If this constraint did not hold, a bidder would not be willing to participate in the auction.

In designing a protocol, we impose the constraints of incentive compatibility, which means that truth-telling is a best policy for each bidder. The revelation principle states that imposing an incentive compatibility constraint does not diminish any generality of discussion [16]. That is, if no desirable mechanism exists on the condition that incentive compatibility holds, no mechanism satisfying the designer's requests exists in the entire design space. This can be explained as follows. Suppose a mechanism has an equilibrium strategy that gives a bidder the highest utility. Bidders participate in the auction by using their proxy agents to carry out the equilibrium strategy. In this case, bidders should report their true valuations of the goods to their proxy agents because it gives them the highest utility. Here, if we consider a new mechanism that includes the original mechanism and the proxy agents, a bidder's best strategy is to report his/her true valuation of the goods when the new mechanism is used.

The incentive compatibility constraint helps to reduce the search space of protocols. In addition, in computational environments, if this constraint holds, a bidder does not have to spy on other bidder valuations, and so the bidder benefits from telling his/her true valuation to the auctioneer, which leads to system stability that simplifies the implementation of a computational agent.

Budget balance means that the seller's revenue coincides with the sum of all bidders' payments. Some readers may think this is trivial. However, in some mechanisms the auctioneer expects to obtain the difference between the bidders' payments and the seller's revenue or cover the deficit, i.e., the difference between the seller's revenue and the sum of all bidders' payments so that the incentive compatibility constraint holds. In the latter case, this problem is more serious. Note that even in the former case, redistributing the difference to the participants spoils incentive compatibility.

2.4 Equilibrium

Given the preference and the constraints, equilibrium should then be examined. Agents are assumed to be self-interested, which makes it difficult to anticipate the outcome we can obtain. In game theory, equilibrium is examined to anticipate the outcome of a developed mechanism. Equilibrium means the stable point in the system, and the properties of the mechanism, such as social surplus and seller's revenue, are discussed in terms of equilibrium. There are several concepts of equilibrium, with dominant strategy equilibrium and Bayesian-Nash equilibrium being representative ones.

Dominant strategy equilibrium means that for any bidder, following the equilibrium strategy is best regardless of the other bidders' strategies, i.e., it gives the maximum utility to each bidder. Bayesian-Nash equilibrium is a weaker concept than dominant strategy equilibrium. Under the assumption that all of the bidders share common beliefs, e.g., in terms of the distribution of valuation values, a bidder gains the most advantage by following the equilibrium strategy if the other bidders also follow it.

If a mechanism can be implemented as dominant strategy equilibrium, a bidder does not have to know the other bidders' valuations, i.e., spying on other bidders' valuation does not benefit him/her. On the other hand, in Bayesian-Nash implementation, if a bidder has a different belief from those of the other bidders, the equilibrium may not be realized. Therefore, dominant strategy implementation is more desirable, especially in network environments. However, dominant strategy implementation is often difficult. That is, there is a trade-off relation between the strength of the equilibrium concept and how wide a range of problems it can deal with.

3 Model

This section presents a formal model to facilitate rigorous discussion. In this paper, we extend the mechanism for auctions in dynamic environments proposed in [6] to the case of auctions including insincere sellers. In the model, accomplishing an exchange of goods can be viewed as a probabilistic event. Therefore, we first describe a model of auctions in dynamic environments and then explain the extended parts.

3.1 Model for auctions in dynamic environments

In auctions in dynamic environments, the benefit from having a good derives not only from using the good but also from whether it is allocated far in advance or just before its use. In other words, the former can be viewed as the primary benefit obtained as a result of using the good, while the latter can be viewed as the secondary benefit obtained during the interval between an allocation determination and the use of the allocated good.

We assume the following.

- We assume a private value model, i.e., *buyer* i's valuation of goods is independent of those of the other buyers.
- For *buyer* i, the valuation is given as follows.

$$v_i() = v_i^P() + v_i^S()$$

v^P represents the benefit obtained as a result of using the allocated good, while v^S represents the benefit obtained during the interval between an allocation determination and the use of the allocated good.
v_i^S might represent disutility because we examine the possibility of re-allocation of goods in this paper. For example, if a seminar organizer wins a room reservation auction, then unfortunately the place of the seminar would change and the organizer would have to inform the seminar participants of this change, which incurs some cost. Thus, we call $v_i^S()$ disutility caused by re-allocation. Note that the organizer can hand a room reservation over to another buyer if the reserved date has not come yet, while the cost for announcing the seminar is a sunk cost.
- v_i^P depends on the environmental conditions when the allocated goods are actually used by the winner.
- The environmental conditions are represented as a set of random variables, $\{cond\} = \{cond_1, cond_2, \cdots, cond_l\}$.
- The domain of a random variable is given and an auctioneer and all buyers know the probability distribution p of each random variable.
For example, the domain of a random variable of weather is {fine, rainy}, and a weather report is shared among the auctioneer and all buyers.
- Thus, the valuation v_i^P of *buyer* i is represented as follows.
Let G denote an allocation of goods and G_i denote allocated goods to *buyer* i in G.
For *buyer* i, the valuation for a bundle of goods G_i is denoted by $v_i^P = v_i^P(G_i; cond_1, cond_2, \cdots, cond_l)$.
We use a notation of $v_i^P(G_i)$ as $v_i^P(G_i; cond_1, cond_2, \cdots, cond_l)$ if there is no confusion.
- We focus on the effect of re-allocation in this paper. Thus, disutility v_i^S of *buyer* i is represented as follows.
For *buyer* i, v_i^S is denoted by $v_i^S(G_i^{prev}, G_i^{current})$. G_i^{prev} represents a bundle of goods previously allocated to *buyer* i, and $G_i^{current}$ represents *buyer* i's current allocation of goods.
v_i^S might be a function of time, e.g., an elapsed time or a remaining time.
- Figure 1 shows an example of a buyer's valuation v_i^P. Here, $\{cond\} = \{weather\}$ and its domain is $\{fine, rainy\}$.
- Figure 2 shows an example of disutility v_i^S caused by the re-allocation itself.
$null$ means that no good is allocated. The disutility caused by changing $null$ to $g1$ represents the disutility that $g1$ is not allocated in advance but just before its use, and thus the time available for preparing oneself to use the allocated good is insufficient. The disutility caused by changing $g1$ to $null$ represents another

	$g1$	$g2$	$(g1, g2)$
buyer 1	10	2	10
buyer 2	4	6	6
buyer 3	3	4	4

(a) fine

	$g1$	$g2$	$(g1, g2)$
buyer 1	5	6	6
buyer 2	7	5	7
buyer 3	6	1	6

(b) rainy

Fig. 1. Valuations of goods conditioned on weather

		TO			
		null	$g1$	$g2$	$(g1, g2)$
	null	0	-1	-1	-1
FROM	$g1$	-3	0	-1	0
	$g2$	-3	-1	0	0
	$(g1, g2)$	-3	0	0	0

Fig. 2. Disutility caused by re-allocation itself

kindx of disutility, e.g., the cost of announcing that a seminar has to be canceled because a seminar room ($g1$) is no longer available. Introducing these disutilities into the discussion is a characteristic of this paper's approach.

- The above representation means that the total valuation of goods is determined by the final allocation of the goods and the environmental conditions, and (dis-)utility accumulates corresponding to the re-allocation from the initial allocation to the final allocation. This enables us to easily describe various problem domains, although it might be possible to assume another functional form of valuations.

3.2 Extension to cases involving insincere sellers

In this section, we explain the extended part of the original model.

- An *auctioneer* exists.
- m *sellers* j exist ($j = 1, \cdots, m$).
- Each *seller* j has one unit or more of the same good to sell, $g_{j1}, g_{j2}, \cdots, g_{jl}$. Different sellers may have different numbers of units.
- For any *seller* j, his/her valuation of the good is 0.

- Each seller is characterized as one of two types: sincere or insincere. A sincere seller completes the exchange of goods and money, while an insincere seller defects in the middle of the process of exchanging goods and money.
- The ratio of sincere sellers to all sellers is $P(sin)$, which is common knowledge. We assume that this value is obtained from the statistical data of past transactions.
- A seller's action is modeled as a probabilistic value. A sincere seller completes the exchange with a probability of $P(suc|sin)$, while an insincere seller completes the exchange with a probability of $P(suc|ins)$.
- Failure may be exogenous as well as the seller's intention. Therefore, $P(suc|sin)$ may be less than one and $P(suc|ins)$ may be more than zero. The former may happen, for example, due to network congestion. The latter may happen because the buyer is satisfied with a good different from that specified in the auction, when the seller sends a different good. Here, $P(suc|sin) > P(suc|ins)$ holds.
- A seller's action is determined only by its type and the probability of $P(suc|sin)$, $P(suc|ins)$. This paper does not consider strategic behaviors of sellers.
  Our model may seem to preclude sellers who perform several successful exchanges only to create a good reputation that he/she then exploits later. Such behavior has been observed on eBay and is very difficult to detect or predict. However, we show that our protocol can mitigate this problem in Section 6.
- *seller* j is characterized by the degree of trust of r_j, which indicates the degree of sincerity of *seller* j. *auctioneer* maintains this value.
- The degree of trust is updated by using the following Bayes rule whenever new information on the result of a trade is obtained.

$$r_j = \frac{r_j P(suc|sin)}{r_j P(suc|sin) + (1 - r_j) P(suc|ins)}.$$

  The initial value of r_j is set to $P(sin)$.
- n *buyer* i exists ($i = 1, \cdots, n$).
- Each buyer's demand for goods is one. Its valuation is v_i. This value corresponds to v_i^P in section 3.1 and is conditioned on the results of the exchange, i.e., success or fail. The value of v_i^S in section 3.1 is set to 0.
- After determining the allocation of the goods and the payment, the following exchange process is assumed. First, a buyer who wins an allocation of the good sends payment to the seller, and then the seller sends the good to the buyer. This means that sellers are tempted to defect in the middle of the exchange process but not buyers.
- *buyer* i's utility u_i is defined as follows. If *buyer* i wins the allocation of the good and the exchange is completed, u_i equals the difference between its valuation and the payment, i.e., $u_i = v_i - payment_i$; if the seller defects, $u_i = -payment_i$. If *buyer* i loses the auction, $u_i = 0$.
- Social surplus is defined by the sum of the increase in the utilities of sincere sellers and buyers. Note that the utilities of insincere sellers are excluded in calculating social surplus because insincere sellers are not rational.
- Let G denote an allocation of goods and G_i denote the allocated good to *buyer* i in G.

- All buyers have the same seller information.
- A transition path represents the results of a series of exchanges such as $\{g_{11}, g_{21}\}$ = $\{success, fail\}$.

In this model, multiple sellers and buyers exist, so this is a double auction. However, the assumption that a seller's action is determined only by its type and the probability of $P(suc|sin)$, $P(suc|ins)$ makes this problem tractable.

3.3 Allocation plan

To maximize social surplus, i.e., a sum of the seller's and buyers' utilities, we need to examine the transition process from one allocation to another as well as the allocation itself, i.e., we have to deal with an allocation plan. An allocation plan consists of an initial allocation and a set of transition rules that specify what allocation to change when some values of random variables occur. Figure 3 shows an example of an allocation plan. This means that $(g1, g2) = (buyer\ 1, buyer\ 2)$ is an initial allocation that changes to $(g1, g2) = (buyer\ 2, buyer\ 1)$ if it rains. If it is fine, the initial allocation is not changed.

initial allocation: $(g1, g2) = (buyer\ 1, buyer\ 2)$
final allocation:
$(g1, g2) = (buyer\ 1, buyer\ 2)$ if fine
$(g1, g2) = (buyer\ 2, buyer\ 1)$ if rainy

Fig. 3. Example of an allocation plan

In a case involving insincere sellers, the allocation is changed whenever an exchange fails.

Next, we give an expression for the utility of $buyer\ i$, $u_(i)$ as follows.

$$\begin{aligned} u_i() &= v_i(AP) - payment_i(AP) \\ &= v_i^P(G_i^{final}; \{cond\}) + v_i^S(AP) - payment_i(AP) \end{aligned}$$

This is called a quasi-linear utility. AP represents an allocation plan. G_i^{final} represents a bundle of goods allocated to $buyer\ i$ in the final allocation.

A method for determining the amount of payment $payment_i$ is described in the next section. We assume that a primary benefit depends only on the final allocation. The disutility caused by the defect of insincere sellers is summed up along the transition path of allocations.

If an allocation plan of AP is given, we can calculate the expected utility based on the degree of trust.

4 Protocol

In this section, we propose a new protocol that uses an allocation plan, which specifies an initial allocation and a set of revision rules [6]. First, we point out that a simple protocol does not satisfy desirable properties, and then we present our protocol. Before describing our protocol, we make the following observations about desirable properties of auction protocols.

- This paper deals with situations where a seller's identity might be uncertain. Thus, individual rationality in this paper means that a buyer does not suffer any loss in expected utility if he or she adopts an equilibrium strategy.
- In this paper a message space consists of information on the results of the exchange as well as valuations of the goods, although this term generally only refers to valuations of the goods in conventional auction design. Therefore, here, incentive compatibility means that a buyer tells the truth about both aspects.

4.1 Failure of a simple auction protocol

As mentioned above, a simple method for allocating goods fixes allocation at the beginning of trading without changing it, even if new seller information is obtained. In this case, a buyer tells the truth about the results of the exchanges because lying does not benefit him/her. However, fixed allocation results in inefficient allocation.

Example 1. Suppose two sellers, $seller$ 1 and $seller$ 2, and four buyers, $buyer$ 1, $buyer$ 2, $buyer$ 3, and $buyer$ 4. Each seller has a good to be sold, g_{11} and g_{21}. $seller$ 1 is insincere, while $seller$ 2 is sincere. $buyer\ i (i = 1, 2, 3, 4)$'s valuations are 10, 8, 6, and 4, respectively. $P(sin) = 0.8$, $P(suc|sin) = 1.0$, and $P(suc|ins) = 0.0$.

$buyer\ i (i = 1, 2, 3, 4)$'s expected valuations of the good are 8, 6.4, 4.8, and 3.2 because $P(sin) = 0.8$. If the auctioneer employs the Vickrey-Clarke-Groves (VCG) protocol for the expected valuation, $buyer\ i (i = 1, 2, 3, 4)$ declares expected valuations of 8, 6.4, 4.8, and 3.2, respectively. Allocation is determined as $(g_{11}, g_{21}) = (buyer\ 1, buyer\ 2)$, and the payments of $buyer$ 1 and $buyer$ 2 are each 4.8. Here, $buyer$ 1 fails to complete the exchange. Based on the exchange results between $buyer$ 1 and $seller$ 1, if the auctioneer changes the allocation of g_{21} from $buyer$ 2 to $buyer$ 1, the expected social surplus increases more than the fixed allocation. However, fixed allocation protocols fail to find this opportunity.

In addition, simply incorporating information on the results of exchanges, i.e., seller reputation, into the protocol and canceling the remainder of exchanges, including those of the insincere seller $seller$ 1, causes another problem. If $seller$ 1 has more than one good to be sold, buyers might be motivated to lie about valuations of the good to avoid becoming the first buyer, since the first buyer may suffer a loss, and this leads to inefficient allocation.

4.2 General protocol for auctions in dynamic environments

Before proposing a protocol for cases involving insincere sellers, we explain a general auction protocol in dynamic environments that re-allocates the goods in response to environmental change, which corresponds to the model described in section 3.1 [6].

An extension of the message space and construction of an allocation plan are novel, although the payment calculation is based on the Vickery-Clarke-Groves (VCG) protocol.

First, we explain a method for constructing an allocation plan that is used to determine an allocation and a payment. It requires enormous computation if we simply enumerate all cases. Thus, we take a method based on dynamic programming.

Here, we assume that there is an order in which values of random variables turn out, specifically such a sequence $cond_1 < cond_2 < \cdots < cond_l$, where first the value of $cond_1$ is produced.

The construction method of an allocation plan is as follows.

1. Buyers declare valuations of any bundle of goods in any combination of values of a random variable to the auctioneer. These valuation values may be true or false.
2. The auctioneer enumerates a combination of buyers' bids so that the allocation feasibility is satisfied, i.e., the same good is not allocated to different buyers simultaneously.
3. The auctioneer calculates social surplus in each case.
4. Next, the auctioneer examines the state before a value of the random variable, $cond_l$, turns out. For each possible allocation, we find the optimal transition rule. Here, optimal means that it maximizes the expected social surplus. Note that possible allocations include allocations in which some of the goods are not allocated to any buyer.
5. Next, the auctioneer considers the state before the values of random variables, $cond_{l-1}$ and $cond_l$ turn out. For each possible allocation, the auctioneer finds the optimal transition rule.
6. The auctioneer continues the above steps until reaching the state before all values of the random variables turn out. At this point, a set of allocation plans, each of which consists of an initial allocation and a set of transition rules, is constructed.
7. The auctioneer finds the allocation plan of AP^* that maximizes the expected social surplus. If two or more allocation plans maximize the expected social surplus, the tie is randomly broken.
8. The auctioneer imposes the following payment on buyers.

$$payment_i = \sum_{j} v_j(AP^*_{-i}) - \sum_{j \neq i} v_j(AP^*)$$

   Here, AP^* represents the optimal allocation plan and AP^*_{-i} represents the optimal allocation plan when *buyer* i does not exist. The amount of $payment_i$ is equal to the other buyers' decrease in the expected valuations by *buyer* i's participation.

Next, we explain how to execute an allocation plan. The execution method of an allocation plan is as follows.

1. Given an allocation plan, the auctioneer announces it to buyers and then sets the initial allocation in the allocation plan to the current allocation.

2. If the value of random variable $cond_1$ turns out, the auctioneer changes the current allocation to another allocation specified in the allocation plan and announces it to the buyers.
3. After the values of all random variables turn out, the final allocation is chosen and the winners consume the allocated goods.

4.3 Protocol for cases involving insincere sellers

Based on the protocol described in the previous section, we propose an auction protocol for cases involving insincere sellers that re-allocates goods in response to the results of previous exchanges. Here, we outline the proposed protocol. It is the same as the protocol in the previous section in that it consists of two phases of constructing an allocation plan and making a payment schedule. However, constructing an allocation plan becomes easier by utilizing the problem structure. A detailed explanation for constructing an allocation plan and making a payment schedule is given below.

1. Buyers declare valuations of the good to the auctioneer. These valuation values may be true or false.
2. The auctioneer finds the allocation plan of AP^* that maximizes the expected social surplus. If more than one allocation plans maximizes the expected social surplus, the tie is randomly broken.
3. The auctioneer imposes payment on buyers.
4. According to the allocation plan and payment schedule, exchanges are carried out.
5. A buyer listed in the allocation plan sends a payment to the seller and reports to the auctioneer whether he/she received the good from the seller.
6. In response to the report from the buyer, the auctioneer changes the allocation as specified in the allocation plan.
7. Allocation is finished if all exchanges are carried out.

This protocol utilizes information on the results of exchanges obtained in the allocation process and allocates goods one by one. However, the protocol is not a sequential auction because valuations of the goods are collected at the beginning of the auction.

4.4 Method for constructing an allocation plan

This section explains how to construct an allocation plan. First, our setting differs from an ordinary auction setting because insincere sellers exist and payments to insincere sellers are not included in the social surplus calculations. Accordingly, we have to clarify what needs to be optimized. Here, the following proposition holds with regard to step 2 in the proposed protocol.

Proposition 1. *If the auctioneer determines the amount of payments so that individual rationality and incentive compatibility are satisfied, the auctioneer's optimization problem can be solved by finding an efficient allocation.*

Proof. The auctioneer's optimization problem is described as follows.

$$\max_{AP} \sum_i (v_i(AP) - payment'_i).$$

Here, AP represents an allocation plan and each $v_i(AP)$ is totaled along the transition path of allocations. $payment'_i$ represents payment to insincere sellers.

If a payment is set to 1 cent, the auctioneer can reduce the amount of loss in the case of insincere sellers, which reduces the decrease of social surplus. However, such a small payment cannot prevent other buyers from overstating their valuations of the good and obtaining the good, which fails to obtain an equilibrium. Therefore, the price should be competitive.

Another possibility for reducing the payment to insincere sellers is to manipulate the number of sales, e.g., to sell only one unit even if sellers have more than one unit to be sold. However, if payment is determined so that individual rationality is satisfied, the following relation holds:

$$v_i(AP) - payment'_i \geq v_i(AP) - payment_i > 0.$$

This means that selling an additional good increases the social surplus.

Therefore, the auctioneer does not have to make excessive efforts to reduce the payment to insincere sellers, i.e., the auctioneer's optimization problem can be solved by finding an efficient allocation. □

Next, we examine the order of exchanges, which determines the amount of social surplus. The following proposition holds in terms of the optimal allocation plan.

Proposition 2. *An allocation plan that allocates a good to a buyer with a higher valuation of the good before allocating it to buyers with lower valuations of the good is optimal for social surplus.*

Proof. Suppose that two goods, g_1 and g_2, and two buyers, $buyer\ H$ and $buyer\ L$, exist. $buyer\ H$ has a higher valuation than $buyer\ L$. Four cases are possible: (1) both exchanges succeed; (2) the first exchange succeeds but the second exchange fails; (3) the first exchange fails but the second exchange succeeds; and (4) both exchanges fail. In cases (1) and (4), the order of exchange does not make any difference, i.e., the social surplus from $(g_1, buyer\ H) \rightarrow (g_2, buyer\ L)$ equals that from $(g_1, buyer\ L) \rightarrow (g_2, buyer\ H)$.

In the case of (2), if $buyer\ L$ first carries out the exchange, $buyer\ H$ cannot win the good, i.e., social surplus is reduced compared to the case where the good is awarded to $buyer\ H$. In (3) when $buyer\ H$ fails the first exchange for g_1, if $buyer\ H$ also carries out the exchange for g_2, the social surplus can be maximized.

Other cases where more than two goods and/or more than two buyers exist can be discussed in the same manner. □

From the above proposition and the assumptions that the same goods are sold and that each buyer's demand for the goods is one, we can conclude that if k exchanges succeed, from the highest to the k-th highest buyers should be included in the k exchanges. Therefore, an allocation plan is obtained as follows.

Initial allocation:

1. Place each buyer in descending order in terms of the valuation of the good: $buyer\ 1, buyer\ 2, \cdots, buyer\ n$.
2. Place each seller in random order: $seller\ 1, seller\ 2, \cdots, seller\ m$.
3. Allocate the goods sold by $seller\ 1$ to $buyer\ 1, \cdots, buyer\ l_1$, where l_1 is the number of goods sold by $seller\ 1$.
4. Allocate the goods sold by $seller\ 2$ to $buyer\ l_1 + 1, \cdots, buyer\ l_1 + l_2$, where l_2 is the number of goods sold by $seller\ 2$.
5. Continue the above steps until all goods are allocated.

Revision rule:

1. If the exchange between $seller\ 1$ and $buyer\ 1$ fails, invoke the above 'Initial allocation' procedure for calculating initial allocation on the condition that $seller\ 1$ has $l_1 - 1$ goods to be sold.
2. If the exchange between $seller\ 1$ and $buyer\ 2$ fails, invoke the above 'Initial allocation' procedure for calculating the initial allocation on the condition that $seller\ 1$ has $l_1 - 1$ goods to be sold.
3. Continue the above steps until all transition paths are examined.

Example 2. Two sellers, $seller\ j(j = 1, 2)$, have goods g_{11} and g_{21}, respectively. The buyers, $buyer\ i(i = 1, 2, 3, 4)$, have valuations of 10, 8, 6, and 4, respectively.

Suppose that g_{11} is chosen as the first good. Here, the choice of the first good can be randomly chosen regardless of the degree of trust. In this case, the following allocation plan is obtained.

initial allocation: $(g_{11}, buyer\ 1) \rightarrow (g_{21}, buyer\ 2)$
revision rule:
$(g_{21}, buyer\ 1)$ if the exchange of $(g_{11}, buyer\ 1)$ fails.

Example 3. $seller\ 1$, has two goods, g_{11}, g_{12}. $buyer\ i(i = 1, 2, 3, 4)$, have valuations of 10, 8, 6, and 4, respectively. In this case, the following allocation plan is obtained.

initial allocation: $(g_{11}, buyer\ 1) \rightarrow (g_{12}, buyer\ 2)$
revision rule:
$(g_{12}, buyer\ 1)$ if the exchange of $(g_{11}, buyer\ 1)$ fails.

Note that an allocation plan is, after all, a plan. If a good is successfully delivered to the buyer, he/she does not have to return it to the auctioneer whenever revision rules are applied after the transaction. This holds from proposition 2.

4.5 Method for calculating payment

This section explains how to calculate payment amount and schedule. A payment schedule is specified when $buyer\ i$ sends its payment to the seller.

$payment_i$ should be determined to induce $buyer\ i$ to tell his/her true valuation of the good. Thus, we employ VCG payment. In each combination of the results of all exchanges, calculate the following:

$$payment_i = \sum_j v_j(AP^*_{-i}) - \sum_{j \neq i} v_j(AP^*).$$

Here, AP^* represents the optimal allocation plan and AP^*_{-i} represents the optimal allocation plan when $buyer\ i$ does not participate in this auction. The amount of $payment_i$ equals the other buyers' decrease in their expected valuations by $buyer\ i$'s participation.

In our problem setting, the auctioneer has to induce buyers to tell true information about the exchange results as well as their valuations of the goods. To satisfy this requirement, payment to each seller is determined so that the following constraints are satisfied.

- Payment to g_j should not be conditioned by the exchange result including g_j.
- Suppose that a buyer purchases a good from $seller\ j$. If the exchange fails, then the buyer purchases a good from $seller\ j'$. In this case, the payment to $seller\ j'$ is conditioned by the exchange results between $seller\ j$ and the buyer.

The former constraint is imposed to prevent buyers from defecting in the middle of the exchange process. Suppose that a buyer sends one cent to the seller, and then the seller sends a good to the buyer, and then the buyer sends the remainder of the payment to the seller. In this case, the buyer is not motivated to send the remainder after receiving the good. Therefore, the remainder should be set to 0.

The latter constraint is imposed to induce buyers to tell the truth about exchange results. That is, payment to $seller\ j'$ prevents the buyer from reporting failure, although the exchange between $seller\ j$ and the buyer actually succeeds.

Here, in an AP allocation plan, several final allocations, $(G_1, G_2, \cdots, G_k)$ exist for the corresponding transition paths, and each final allocation is realized with probability $p(G)$. That is, $v_j(AP) = \sum_l p(G_l) v_j(G_l)$. Thus, the auctioneer decomposes $payment_i$ into parts corresponding to each final allocation and arranges a payment schedule to satisfy the above constraints. From this discussion, payments are calculated as follows.

Payment calculation:

1. For each final allocation, i.e., each transition path, calculate payment based on the above expression of $payment_i$, where the degree of trust r_j is updated according to the results of preceding exchanges.
2. For each buyer and good, add up the payments obtained in step 1 weighted with the probability of each transition path.
3. Exchanges are carried out one by one. Each payment is conditioned by the results of previous exchanges.

The details of this payment calculation are illustrated in the following two examples.

Example 4. Two sincere sellers, $sellers\ j(j = 1, 2)$, have goods, g_{11} and g_{21}, respectively. Four buyers, $buyer\ i(i = 1, 2, 3, 4)$, have valuations of 10, 8, 6, and 4, respectively. In addition, the ratio of sincere sellers to all sellers $P(sin)$ is 0.8, no history of either seller's transactions is available, and $P(suc|sin) = 1.0$ and $P(suc|ins) = 0.0$. In this case, the obtained allocation plan is represented in Example 2, and payments are calculated as follows.

$buyer$ 1:

- If the exchanges of g_{11} and g_{21} both succeed, the payment to g_{11} is $(8+6)-8 = 6$, and the payment to g_{21} is 0.
- If the exchange of g_{11} succeeds while the exchange of g_{21} fails, the payment to g_{11} is $8 - 0 = 8$, and the payment to g_{21} is 0.
- If the exchange of g_{11} fails while the exchange of g_{21} succeeds, the payment to g_{11} is 0, and the payment to g_{21} is $8 - 0 = 8$.
- If the exchanges of g_{11} and g_{21} both fail, the payments to g_{11} and g_{21} are $0 - 0 = 0$

Therefore, the payment to $seller$ 1 is calculated as $0.8 \times 0.8 \times 6 + 0.8 \times 0.2 \times 8 = 5.12$. The payment of 5.12 is paid to $seller$ 1 at the beginning of the exchange process between $buyer$ 1 and $seller$ 1.

On the other hand, the payment to $seller$ 2 is calculated as $0.8 \times 8 = 6.4$ when the exchange of g_{11} fails. Therefore, the payment to $seller$ 2 is 0 if the exchange of g_{11} succeeds, while the payment to $seller$ 2 is 6.4 if it fails.

$buyer$ 2:

- If the exchanges of g_{11} and g_{21} both succeed, the payment to g_{11} is 0, and the payment to g_{21} is $(10 + 6) - 10 = 6$.
- If the exchange of g_{11} succeeds while the exchange of g_{21} fails, the payments to g_{11} and g_{21} are $0 - 0 = 0$.
- If the exchange of g_{11} fails while the exchange of g_{21} succeeds, the payments to g_{11} and g_{21} are $0 - 0 = 0$.
- If the exchanges of g_{11} and g_{21} both fail, the payments to g_{11} and g_{21} are $0 - 0 = 0$

Therefore, $buyer$ 2 sends a payment of $0.8 \times 6 = 4.8$ to $seller$ 2 at the beginning of the exchange between $buyer$ 2 and $seller$ 2 if the exchange between $buyer$ 1 and $seller$ 1 succeeds, while the payment by $buyer$ 2 is 0 if the exchange between $buyer$ 1 and $seller$ 1 fails.

The expected utility of $buyer$ 1 is $0.8\times0.8\times10+0.8\times0.2\times10+0.2\times0.8\times10-5.12-0.2\times6.4 = 3.2$. The expected utility of $buyer$ 2 is $0.8\times0.8\times8-0.8\times4.8 = 1.28$. The revenues of $seller$ 1 and $seller$ 2 are 5.12, $0.2 \times 6.4 + 0.8 \times 4.8 = 5.12$, respectively.

Example 5. Suppose a sincere seller, $seller$ 1, sells two units of the good, g_{11} and g_{12}. $buyer\ i(i = 1, 2, 3, 4)$'s valuations are 10, 8, 6, and 4, respectively. $P(sin) = 0.8$, $P(suc|sin) = 0.98$, and $P(suc|ins) = 0.02$.

For g_{12}, if the exchange of g_{11} succeeds, the degree of trust r_1 is updated from 0.8 to 0.995; if the exchange of g_{11} fails, r_1 is updated from 0.8 to 0.005. In this case, the obtained allocation plan is represented in Example 3, and the payments are calculated as follows.

buyer 1:

- If the exchanges of g_{11} and g_{12} both succeed, the payment to g_{11} is $(8+6)–8=6$, and the payment to g_{12} is 0.
- If the exchange of g_{11} succeeds while the exchange of g_{21} fails, the payment to g_{11} is $8-0=8$, and the payment to g_{12} is 0.
- If the exchange of g_{11} fails while the exchange of g_{21} succeeds, the payment to g_{11} is 0, and the payment to g_{12} is $8-0=8$.
- If the exchanges of g_{11} and g_{21} both fail, the payments to g_{11} and g_{12} are $0-0=0$

Therefore, the payment to g_{11} is calculated as $(0.8\times0.98+0.2\times0.02)\times(0.995\times 0.98+0.005\times0.02)\times6+(0.8\times0.98+0.2\times0.02)\times(0.005\times0.98+0.995\times 0.02)\times8=4.7670848$. A payment of 4.7670848 is paid to *seller* 1 at the beginning of the exchange of g_{11}. In addition, the payment to g_{12} is calculated as $(0.005\times 0.98+0.995\times0.02)\times8=0.1984$. The payment of 0.1984 is paid to *seller* 1 at the beginning of the exchange of g_{12} if the exchange of g_{11} fails.

buyer 2:

- If the exchanges of g_{11} and g_{12} both succeed, the payment to g_{11} is 0, and the payment to g_{12} is $(10+6)-10=6$.
- If the exchange of g_{11} succeeds while the exchange of g_{12} fails, the payments to g_{11} and g_{12} are $0-0=0$.
- If the exchange of g_{11} fails while the exchange of g_{12} succeeds, the payments to g_{11} and g_{12} are $0-0=0$.
- If the exchanges of g_{11} and g_{12} both fail, the payments to g_{11} and g_{12} are $0-0=0$

Therefore, *buyer* 2 sends a payment of $(0.995\times0.98+0.005\times0.02)\times6=5.8512$ to *seller* 1 at the beginning of the exchange of g_{12} if the exchange between *buyer* 1 and *seller* 1 succeeds.

The expected utility of *buyer* 1 is $(0.8\times0.98+0.2\times0.02)\times(0.995\times0.98+ 0.005\times0.02)\times10+(0.8\times0.98+0.2\times0.02)\times(0.005\times0.98+0.995\times0.02)\times 10+(0.2\times0.98+0.8\times0.02)\times(0.005\times0.98+0.995\times0.02)\times10-4.7670848- (0.2\times0.98+0.8\times0.02)\times0.1984=3.1234304$. The expected utility of *buyer* 2 is $(0.8\times0.98+0.2\times0.02)\times(0.995\times0.98+0.005\times0.02)\times8-(0.8\times0.995+ 0.2\times0.02)\times5.8512=1.5369152$. The revenues of *seller* 1 are $4.767084+0.02\times 0.1984+0.98\times5.8512=10.505228$.

5 Properties of the proposed protocol

This section proves that the proposed protocol satisfies desirable properties.

We use the following two lemmas to prove that the proposed protocol satisfies incentive compatibility.

Lemma 1. *For each buyer, truth telling about valuations of the good is a best policy if the buyer tells the truth about the results of the exchange.*

Proof. If the buyer tells the truth about the exchange results, the expected utility of each buyer is calculated as follows.

$$\begin{aligned} u_i &= v_i(AP^*) - payment_i(AP^*) \\ &= v_i(AP^*) - (\sum_j v_j(AP^*_{-i}) - \sum_{j \neq i} v_j(AP^*)) \\ &= v_i(AP^*) + \sum_{j \neq i} v_j(AP^*) - \sum_j v_j(AP^*_{-i}). \end{aligned}$$

By inspecting the above expression, we can see that the third term is independent from *buyer* i's declaration, i.e., *buyer* i cannot manipulate it. The first and second terms equal the objective function that an auctioneer tries to maximize. Therefore, for *buyer* i, truth telling about valuations of the goods is a best strategy if the buyer tells the truth about the exchange results. □

Lemma 2. *For each buyer, truth telling about the results of exchanges is a best policy if the buyer tells the truth about the valuations of the good.*

Proof. If a buyer lies about the results of exchanges, i.e., reports success when the exchange fails, the buyer loses subsequent exchange opportunities. Consider a subplan starting when the exchange fails. Because a buyer with a higher valuation wins the good before buyers with lower valuations, the utility obtained in each subplan of the allocation plan is larger than or equal to zero. This is because the participation of the buyer with the highest valuation increases social surplus. Thus, a false declaration that the exchange succeeds does not increase buyer utility.

On the other hand, if a buyer lies about the results of exchanges, i.e., reports failure when the exchange succeeds, the buyer is required to carry out the next exchange, which incurs a payment. Because the buyer's demand is assumed to be one, the utility from obtaining the second good is less than zero. Therefore, a false declaration that the exchange fails does not increase the buyer's utility. □

Proposition 3. *For each buyer, truth telling about both the valuation of the good and the results of exchanges is a best policy in the proposed protocol.*

Proof. From lemmas 1 and 2, we can directly obtain this proposition. □

Proposition 4. *The proposed protocol can attain an allocation that maximizes the expected social surplus.*

Proof. The proposed protocol can induce buyers to declare true valuations and find such an allocation plan that maximizes the expected social surplus. Thus, the proposed protocol attains an allocation that maximizes the expected social surplus. □

Proposition 5. *The proposed protocol can attain an ex post Pareto efficient allocation.*

Proof. As proved in Proposition 2, a selected allocation plan is a scheme that allocates a good to the buyer with the higher valuation of the good before allocating it to buyers with lower valuations of the good. Therefore, if l exchanges succeed in the end, the buyers from the highest to the l-th highest valuation of the good win the good. □

Proposition 6. *The proposed protocol satisfies the individual rationality constraint of expected values.*

Proof. If the participation of a buyer improves social surplus, the buyer's bid is included in an allocation plan, and the buyer's utility equals the increase of social surplus based on his/her participation. If the participation of a buyer does not improve social surplus, the buyer's bid is not included in an allocation plan, and the buyer's payment is zero. Therefore, the utility of each buyer is larger than or equal to zero. □

Note that the proposition holds for expected values. It may happen that a buyer's utility becomes less than zero in some cases, although the expected utility is larger than or equal to zero.

6 Discussion

As mentioned above, sellers sometimes repeatedly perform several successful exchanges merely to create a good reputation that he/she can then exploit later. Such serious fraud has been observed on eBay and is very difficult to detect or predict. Such a seller can be viewed as rational, i.e., behaving to maximize his or her utility, while insincere sellers in this paper can be considered irrational. This means that when the proposed protocol is employed, if a seller is rational and can correctly deliver the good, defecting in the middle of the process of exchanging goods and money reduces the seller's utility.

Suppose the same setting as Example 5. If $seller$ 1 completes the two exchanges of goods and money, the expected revenues of $seller$ 1 are $4.7670848 + 0.98 \times 5.8512 + 0.02 \times 0.1984 = 10.505228$. On the other hand, if $seller$ 1 intentionally fails the first exchange, the expected revenues of $seller$ 1 are $4.7670848 + 0.02 \times 5.8512 + 0.98 \times 0.1984 = 5.0785408$. That is, intentional failure heavily reduces revenue. Thus, rational sellers are not motivated to commit fraud. Therefore, the proposed protocol can mitigate the problem of performing several successful exchanges only to create a good reputation to be exploited later.

As a limitation of the proposed protocol, it requires that all goods be known before the auction starts and does not allow a new seller to participate after the auction starts.

Another limitation of the proposed protocol is the computational burden. Sandholm and Lesser pointed out the problems of contingency contracts; there is a potential combinatorial explosion of goods that exchanges are conditioned on, and it is often impossible to enumerate all possible relevant future events in advance [24]. In addition, the VCG protocol needs to solve a combinatorial optimization problem when calculating an allocation and a payment.

However, our situation does not so heavily suffer in this regard because we assume that each buyer's demand for goods is one. From proposition 2, we can easily find revision rules; if the number of goods is l, the buyers from the highest to the l-th highest valuations are included in the final allocation.

In the payment calculation, at most the auctioneer has to examine 2^l cases for each buyer. In Example 4, we examined four cases where: (1) the exchanges of g_{11} and g_{21} both succeed; (2) the exchange of g_{11} succeeds while the exchange of g_{21} fails; (3) the exchange of g_{11} fails while the exchange of g_{21} succeeds; and (4) the exchanges of g_{11} and g_{21} both fail. Developing an efficient payment calculation method is part of our future work.

Finally, the proposed protocol for when and how much a buyer should pay seems complicated for humans but not for computational agents. A drawback of the proposed protocol is that exchanges must be carried out one by one, which requires more time to complete them. Mitigating this problem is also included in our future work.

7 Conclusions

This paper proposed an auction protocol for a multi-unit single-item auction for multiple sellers, some of whom may be insincere sellers. This protocol creates an allocation plan that specifies an initial plan and revision rules based on the results of preceding exchanges. The proposed protocol can find an allocation that maximizes the expected social surplus. In this paper, we (1) presented a method for constructing an allocation plan and making a payment schedule and (2) showed that the proposed protocol can induce buyers to declare true valuations and true information about exchange results.

In some cases, a buyer obtains disutility when the exchange fails, e.g., the buyer is a contractee of some task and has to pay a penalty if its task completion is delayed. In this case, buying a good from a seller identified as sincere is valuable. However, this may provoke a situation where some buyer's investigation of an unidentified seller considerably improves social surplus, which makes the investigator's payment a large negative value. This may cause the auctioneer to engage in deficit spending. A similar problem is reported in [3]. Solving this problem will be tackled in our future work.

References

1. Dash, R.K., Jennings, N.R., Parkes, D.C.: Computational-mechanism design: a call to arms. IEEE Intelligent Systems **18**(6) (2003) 40–47

2. Yokoo, M., Sakurai, Y., Matsubara, S.: The effect of false-name bids in combinatorial auctions: New fraud in internet auctions. Games and Economic Behavior **46**(1) (2004) 174–188
3. Ito, T., Yokoo, M., Matsubara, S.: Designing an auction protocol under asymmetric information on nature's selection. In: Proceedings of the First International Joint Conference on Autonomous Agents and Multiagent Systems (AAMAS-2002). (2002)
4. Ito, T., Yokoo, M., Matsubara, S.: A combinatorial auction among versatile experts and amateurs. In: Proceedings of the third International Joint Conference on Autonomous Agents and Multiagent Systems (AAMAS-2004). (2004) 378–385
5. Porter, R., Ronen, A., Shoham, Y., Tennenholtz, M.: Mechanism design with execution uncertainty. In: Proceedings of the Eighteenth Conference on Uncertainty in Artificial Intelligence (UAI-02). (2002)
6. Matsubara, S.: Auction in dynamic environments: Incorporating the cost caused by re-allocation. In: Proceedings of the Fourth International Joint Conference on Autonomous Agents and Multiagent Systems (AAMAS-2005). (2005) 643–649
7. National Consumers League: NCL's online auction survey: Summary (2001) http://www.nclnet.org/shoppingonline/auctionsurvey.htm.
8. Sandholm, T.W., Lesser, V.R.: Equilibrium analysis of the possibilities of unenforced exchange in multiagent systems. In: Proceedings of the Fourteenth International Joint Conference on Artificial Intelligence (IJCAI-95). (1995) 694–701
9. Matsubara, S., Yokoo, M.: Defection-free exchange mechanisms for information good. In: Proceedings of the Fourth International Conference on Multi-Agent Systems (ICMAS-2000). (2000) 183–190
10. Matsubara, S., Yokoo, M.: Defection-free exchange mechanisms based on an entry fee imposition. Artificial Intelligence Journal **142**(2) (2002) 265–286
11. Resnick, P., Zeckhauser, R.: Trust among strangers in internet transactions: Empirical analysis of ebay's reputation system. In Baye, M.R., ed.: The Economics of the Internet and E-Commerce. Volume 11 of Advances in Applied Microeconomics. Elsevier Science (2002)
12. Zacharia, G., Moukas, A., Maes, P.: Collaborative reputation mechanisms in electronic marketplaces. In: Proceedings of the Thirty-second Hawaii International Conference on System Sciences (HICSS-99). (1999)
13. Avery, C., Resnick, P., Zeckhauser, R.: The market for evaluations. American Economic Review **89**(3) (1999) 564–584
14. Tennenholtz, M.: Reputation systems: An axiomatic approach. In: Proceedings of the Twentieth Conference on Uncertainty in Artificial Intelligence (UAI-04). (2004)
15. Dellarocas, C.: Efficiency through feedback-contingent fees and rewards in auction marketplaces with adverse selection and moral hazard. In: Proceedings of the Third ACM Conference on Electronic Commerce (EC-03). (2003) 11–18
16. Mas-Colell, A., Whinston, M.D., Green, J.R.: Microeconomic Theory. Oxford University Press (1995)
17. Parameswaran, M., Susarla, A., Whinston, A.B.: P2P networking: An information-sharing alternative. IEEE Computer **34**(7) (2001) 31–38
18. Gupta, A., Jukic, B., Parameswara, M., Stahl, D.O., Whinston, A.B.: Streamlining the digital economy: how to avert a tragedy of the commons. IEEE Internet Computing **1**(6) (1997) 38–46
19. Feldman, M., Chuang, J., Stoica, I., Shenker, S.: Hidden-action in multi-hop routing. In: Proceedings of the Sixth ACM Conference on Electronic Commerce (EC-05). (2005) 117–126

20. Krishna, V.: Auction Theory. Academic Press (2002)
21. Milgrom, P.: Putting Auction Theorey to Work. Cambridge University Press (2004)
22. Klemperer, P.: Auctions: Theory and Practice. Princeton University Press (2004)
23. Rasmusen, E.: Games and Information: an introduction to game theory. 3rd edn. Blackwell Publishers (2001)
24. Sandholm, T.W., Lesser, V.R.: Advantages of a leveled commitment contracting protocol. In: Proceedings of the Thirteenth National Conference on Artificial Intelligence (AAAI-96). (1996) 126–133

Cheating in Second Price Auctions with Affiliated Values

Takahiro Watanabe[1] and Takehiko Yamato[2]

[1] Faculty of Urban Liberal Arts,
Tokyo Metropolitan University,
1-1 Minami-Osaka, Hachioji, Tokyo 192–0397, Japan,
forward0@nabenavi.net

[2] Department of Value and Decision Science,
and Department of Social Engineering,
Graduate School of Decision Science and Technology,
Tokyo Institute of Technology,
2-12-1 Ookayama, Meguro-ku, Tokyo 152-8552, Japan.

Summary. We model cheating in second price auctions when valuations of bidders are affiliated. There exists a cheating opportunity for an unfair seller who can observe the bids submitted by all bidders and then submits his extra bid to increase the payment of the winning bidder. We derive a symmetric equilibrium strategy when bidders predict this cheating behavior by the seller. We also conduct a numerical analysis to compare a second price auction with a first price auction in terms of expected revenues with the possibility of cheating. The expected revenue of a non-cheating seller from a second price auction is still larger than that from a first price auction if two bidders' belief for cheating are sufficiently small and the correlation between signals is large enough.

1 Introduction

Since Vickrey ([1961]) and Milgrom and Weber ([1982]), auction theory has been developed in the field of game theory. In the auction theory, second price auctions proposed by Vickrey ([1961]) have advantages over first price auctions in several points such as stability for collusion and incentive compatibility attaining efficient allocations. It is especially worth noting that the expected revenue of a seller from a second price auction is not less than one from a first price auction in the symmetric model in which bidders values are affiliated .

However, Rothkopf, Teisberg and Kahn ([1990]) pointed out that second price auctions are rarely used in practice in spite of their advantages. They stressed one of the important reasons for this fact is the possibility of cheating for sellers in second price auctions. They assert that there exists a cheating opportunity for an unfair seller who can observe the bids by all bidders in a second price auction. The seller will submit an extra bid just below the maximum submitted bid in order to sell the object

T. Watanabe and T. Yamato: *Cheating in Second Price Auctions with Affiliated Values*, Studies in Computational Intelligence (SCI) **110**, 61–72 (2008)
www.springerlink.com

at the first price even in the second price auction. If bidders predict the possibility of cheating, then they will shave their bids and will not tell their valuations truthfully. Therefore, with fear of cheating, a second price auction may become less profitable than a first price auction for a non-cheating and fair seller.

Rothkopf and Harstad ([1995]) confirmed these cheating effects in a formal model. They conclude that all sellers except the type with the highest cheating probability strictly prefer first price auctions to second price auctions. Moreover, Porter and Shoham ([2005]) explicitly derived a symmetric equilibrium strategy of each bidder in a second price auction, depending on a given probability of bid-taker's cheating, and then investigated a loss in the expected revenue of a non-cheating seller due to the possibility of cheating.

All of these past studies claimed that a second price auction gives a non-cheating seller a smaller expected revenue than a first price auction does when bidders fear the possibility of cheating. However, it was assumed that the values of bidders are independently distributed. The assumption of independence is restrictive and rules out possibilities of correlations among bidders' values. In this paper, we make a more general assumption that the values of bidders are *affiliated*, including independent values and common values as special cases (Milgrom and Weber (1982)). With independent values, the expected revenue from a second price auction with no possibility of cheating is equivalent to that from a first price auction, so that it is easy to understand that for a non-cheating and fair seller, a second price auction is worse than a first price auction in the existence of fear of cheating. On the other hand, when the values of bidders are affiliated, the expected revenue from a second price auction in the absence of cheating is larger than or equal to that from a first price auction. Then the following question naturally arises: does a second price auction still has a disadvantage to a non-cheating and fair seller in compared with a first price auction even with affiliated values?

In order to investigate this question, we construct a formal model in which all bidders predict cheating behavior of an unfair seller according to some probability. We derive a symmetric equilibrium strategy by solving a differential equation derived as the condition on maximization of an expected utility of a bidder. To examine the expected revenue of a non-cheating seller, we restrict to the case of two bidders in which the value of the object for each bidder is given by the sum of each bidder's signal which is private information of the bidder. We introduce a certain class of distribution functions for the signals of two bidders such that correlation coefficients of the signals depend on one parameter.

We numerically calculate the expected revenue of a non-cheating seller from a second price auction and compare it to that from a first price auction. We show that expected revenue of a non-cheating seller from a second price auction is larger than a first price auction if beliefs of bidders for cheating is sufficiently small and correlation of signals is large enough.

The paper is organized as follows. In section 2, we define our model. In section 3, we derive a symmetric equilibrium strategy. In section 4, we limit our attention to the case of two bidders and define a certain class of distribution functions of bidders' signals which are correlated and affiliated. We conduct a numerical analysis to compare a second price auction with a first price auction in terms of expected revenues with the possibility of cheating.

2 The Model

We consider the following game with one seller and n bidders who want to trade a single object by using a second price auction. The set of bidders is denoted by $N = \{1, \cdots, n\}$. Each bidder i receives a signal x_i which is the realization of the random variable $X_i \in [0, \bar{x}]$. A signal x_i of bidder i means his private information about the value of the object. It is assumed that the value of the object to each bidder is a function of all bidders' signals. We also assume that the function is symmetric and identical to all bidders. Hence, the value of the object for bidder i is expressed by $v(x_i, x_{-i})$ which is symmetric in the last $n-1$ variables, where $x_{-i} = (x_1, \ldots, x_{i-1}, x_{i+1}, \ldots, x_n)$. Let f be the joint probability density of the random variables $X_1, \ldots, X_n$. We assume that f is strictly positive, symmetric and continuous on $[0, \bar{x}]^n$. According to the standard analysis in auction theory for symmetric interdependent values (Milgrom and Weber([1982]), Krishna([2002]), Menezes and Montero ([2005]), we assume that the random variables $(X_1, \ldots, X_n)$ are affiliated. The random variables $(X_1, \ldots, X_n)$ are said to be affiliated if f satisfies

$$f(x \vee x')f(x \wedge x') \geq f(x)f(x')$$

for any $x = (x_1, \ldots, x_n)$ and $x' = (x'_1, \ldots, x'_n)$ where

$$x \vee x' = (\max(x_1, x'_1), \ldots, \max(x_n, x'_n))$$

and

$$x \wedge x' = (\min(x_1, x'_1), \ldots, \min(x_n, x'_n)).$$

There are two possible types for the seller, a cheating seller and a non-cheating seller. The cheating seller can observe all bidders' bids before the auction is finished and submit an extra bid set just below the maximum bid of all bidders'. Thus, since the second price in the auction is almost near the maximum bid and the winner has to pay the amount that almost equals to the first price auction. Finally, the cheating seller can sell the object almost at the first price of all bidders' bids even in the second price auction. On the other hand, such cheating is not allowed for the non-cheating seller and he should sell the object at the second price as an usual way. Each bidder observes his signal of the object $X_i = x_i$, but he knows neither the seller's type nor the other bidders' signals x_{-i}. We assume that all bidders predict that the seller's type will be cheating with the probability $\mu \in [0, 1]$ and this fact is common knowledge. μ is called a belief of bidders for cheating. Each bidder i submits his bid $b_i \in [0, \bar{x}]$ simultaneously, depending on his signal $x_i \in [0, \bar{x}]$ and a belief $\mu \in [0, 1]$. We also assume that each bidder follows a symmetric strategy β_μ when the probability of cheating is μ. Let $\beta_\mu(x_i)$ be the submitting bid of bidder i with his signal x_i and the probability of cheating μ. We assume that $\beta_\mu(x)$ is increasing and differentiable in x.

Now we define each bidder's payoff. Given bidders' signals $(x_1, \ldots, x_n) \in [0, \bar{x}]^n$, $u_i((x_1, \ldots, x_n), b_i, \beta_\mu)$ denotes bidder i's payoff when his bid is $b_i \in [0, \bar{x}]$ and all bidders except i follow the symmetric strategy β_μ. Notice that each bidder who wins an auction pays the second price when the seller is non-cheating, whereas he does his own bid when the seller in cheating. Hence the payoff of each bidder i is expressed as follows:

$$
\begin{aligned}
&u_i((x_1, \ldots, x_n), b_i, \beta_\mu) \\
&= \begin{cases} v(x_1, \ldots, x_n) - \{\mu b_i + (1-\mu) max_{j \neq i} \beta_\mu(x_j)\} & \text{if } b_i > max_{j \neq i} \beta_\mu(x_j), \\ 0 & \text{otherwise.} \end{cases}
\end{aligned}
$$

Note that when $\mu = 0$, each bidder's payoff is equivalent to that in an usual second price auction because he believes no cheating in the auction. In contrast, when $\mu = 1$, the payoff is equivalent to that in a first price auction because each bidder believes that he always pay his own bid if he wins. The case of ties is negligible for our analysis because we assume a continuous joint distribution for signals of bidders and so a probability of ties is zero in equilibria.

Now we define expected revenues for the two types of the seller in a second price auction. Let $U_C(\beta_\mu)$ (resp. $U_{NC}(\beta_\mu)$) be the expected revenue of the cheating (resp. non-cheating) seller when all bidders follow the same belief of cheating μ and the same strategy β_μ. Also, let Z_1 and Z_2 be the highest and the second highest signals among all bidder's signals $(X_1, \ldots, X_n)$. The distribution functions of Z_1 and Z_2 are denoted by H_1 and H_2 respectively, and the corresponding density functions are given by h_1 and h_2. Then the expected revenues $U_C(\beta_\mu)$ and $U_{NC}(\beta_\mu)$ are provided as follows:

$$
U_C(\beta_\mu) = \int_0^{\bar{x}} \beta_\mu(z) h_1(z) dz, \text{ and}
$$

$$
U_{NC}(\beta_\mu) = \int_0^{\bar{x}} \beta_\mu(z) h_2(z) dz.
$$

3 Equilibrium Strategies

In this subsection, we will define symmetric equilibria of our game and derive them. To define the equilibrium concept, we first introduce some notation and definitions. Let X_{-i} be an $(n-1)$ dimensional vector of the random variables among the $n-1$ bidders except bidder i, that is,

$$
X_{-i} = (X_1, \ldots, X_{i-1}, X_{i+1} \ldots, X_n).
$$

Let the random variables $Y_1 = max\{X_1, \ldots, X_{i-1}, X_{i+1} \ldots, X_n\}$ be the highest signal of X_{-i}. Let $G(\cdot|x)$ be the conditional distribution of Y_1 given $X_i = x$. In other words, $G(y|x)$ is the conditional distribution defined by

$$G(y|x) = Prob[Y_1 \leq y|X_i = x].$$

The conditional density function corresponding to $G(y|x)$ is denoted by $g(y|x)$. We define the function $v(x, y)$ which is the expectation of bidder i's value on the condition that his signal is $X_i = x$ and the highest signal among the other bidders equals to y as follows:

$$v(x, y) = E[v(x, X_{-i})|X_i = x, Y_1 = y].$$

We assume that $v'(y, y)$ is bounded for any $y > 0$, that is, there exists $M > 0$ such that $v'(y, y) < M$ for any $y > 0$. Given bidder i's signal $x \in [0, \bar{x}]$, let $U_i(x, b_i, \beta_\mu)$ be his expected payoff when his bid is $b_i \in [0, \bar{x}]$ and all bidders except i follow the same strategy β_μ. It is calculated as follows:

$$U_i(x, b_i, \beta_\mu) = \int_{b_i > \beta_\mu(y)} (v(x, y) - \mu b_i - (1 - \mu)\beta_\mu(y))g(y|x)dy.$$

Now we define a symmetric equilibrium strategy of our game.

Definition 1. *Given the belief of cheating $\mu \in [0, 1]$, β^*_μ is a symmetric equilibrium, or simply an equilibrium strategy if and only if*

$$\begin{array}{l} \forall i \in N, \forall x \in [0, \bar{x}], \forall b_i \in [0, \bar{x}], \\ U_i(x, \beta^*_\mu(x), \beta^*_\mu) \geq U_i(x, b_i, \beta^*_\mu) \end{array}$$

It means that for any valuation x, each bidder i maximizes his conditional expected payoff by choosing his bid according to the equilibrium strategy β^*_μ and all other bidders follow the equilibrium strategy β^*_μ.

Next we derive the equilibrium strategy of each bidder. Let $\Pi(z, x)$ be the expected payoff of bidder i when the other bidders $j \neq i$ follow the equilibrium strategy β_μ, the signal of bidder i is x, and he bids $\beta_\mu(z)$. The expected profit $\Pi(z, x)$ is expressed by

$$\begin{aligned} \Pi(z, x) &= U_i(x, \beta_\mu(z), \beta_\mu) \\ &= \mu \int_{\beta_\mu(z) > \beta_\mu(y)} (v(x, y) - \beta_\mu(z))g(y|x)dy \\ &\quad + (1 - \mu) \int_{\beta_\mu(z) > \beta_\mu(y)} (v(x, y) - \beta_\mu(y))g(y|x)dy \end{aligned}$$

Since β_μ is strictly increasing, we have

$$\Pi(z, x) = \mu \int_0^z (v(x, y) - \beta_\mu(z))g(y|x)dy + (1 - \mu) \int_0^z (v(x, y) - \beta_\mu(y))g(y|x)dy.$$

Differentiating $\Pi(z, x)$ with respect to z and using Leibniz's formula, we obtain

$$\begin{aligned} \frac{\partial}{\partial z} \Pi(x, z) &= \mu(v(x, z)g(z|x) - \beta'_\mu(z)G(z|x) - \beta_\mu(z)g(z|x)) \\ &\quad + (1 - \mu)(v(x, z) - \beta_\mu(z))g(z|x) \\ &= \mu(v(x, z) - \beta_\mu(z))g(z|x) - \mu\beta'_\mu(z)G(z|x). \end{aligned}$$

If β_μ is an equilibrium strategy, then $\Pi(z,x)$ maximizes at $z = x$. Hence, $\frac{\partial \Pi}{\partial z}|_{z=x} = 0$. This implies a differential equation expressed by:

$$\mu\beta'_\mu(x)G(x|x) = (v(x,x) - \beta_\mu(x))g(x|x) \tag{1}$$

When $\mu = 0$, (1) derives $\beta_\mu(x) = v(x,x)$ and we find that it is an equilibrium strategy for an usual second price auction with affiliated values (e.g., see Milgrom and Weber ([1982]) and Krishna ([2002]).

Next we consider $\mu > 0$. If $x = 0$, then $\beta_\mu(0) = 0$ in an equilibrium strategy because a positive bid gives a negative expected payoff. Hence, solving the differential equation (1) by using a boundary condition $\beta_\mu(0) = 0$, we have:

$$\beta_\mu(x) = \int_0^x v(y,y)dL_\mu(y|x),$$

where

$$L_\mu(y|x) = exp\left(-\frac{1}{\mu}\int_y^x \frac{g(t|t)}{G(t|t)}dt\right); \text{ and}$$

The condition (1) is merely a necessary condition for an equilibrium strategy, but it is easily to see that $\beta_\mu(x)$ satisfies a sufficient condition for an equilibrium strategy by a similar discussion like as standard auction theory. In sum, an equilibrium strategy is expressed as the following proposition.

Proposition 1. *Given any belief μ, an equilibrium strategies β^*_μ are provided as follows for any $x \in [0, \bar{x}]$.*

(i) if $\mu > 0$, then

$$\beta^*_\mu(x) = \int_0^x v(y,y)dL_\mu(y|x),$$

where

$$L_\mu(y|x) = exp\left(-\frac{1}{\mu}\int_y^x \frac{g(t|t)}{G(t|t)}dt\right); \text{ and}$$

(ii) if $\mu = 0$, then $\beta^*_\mu(x) = v(x,x)$.

Porter and Shoham ([2005]) derive an equilibrium strategy in a second price auction when values of the object are privately and independently distributed. Our result is an extension of their result when the values of the object are affiliated.

In the following, $E^C(\mu)$ denotes the expected revenue of a cheating seller when each bidder follows an equilibrium strategy β^*_μ for a given belief μ, and $E^{NC}(\mu)$ stands for that of a non-cheating seller:

$$E^C(\mu) = U_C(\beta^*_\mu),\ E^{NC}(\mu) = U_{NC}(\beta^*_\mu).$$

We also denote the expected revenue of a seller for a first price auction by E^1. Note that the expected revenue of a cheating seller from a first price auction is the same as that of a non-cheating seller, and it is independent of the value of belief μ, because observing all bids has no effect in a first price auction.

4 Numerical Analysis

In this section, we restrict our attention to the case of two bidders and give numerical analysis for a certain class of distribution functions on bidders' signals. For this class of distribution functions, the signals are affiliated. In this case, it is well-known that the expected revenue from a second price auction is greater than or equals to that from a first price auction in the standard auction model with no cheating (Milgrom and Weber ([1982])). We calculate and compare the expected revenues from the two auctions with the possibility of cheating, given several bidders' beliefs and parameters used to specify distribution functions.

4.1 A Class of Distribution Functions

We assume that each signal X_i is distributed on $[0,1]$ $(i = 1,2)$. Let us consider the following class of distribution functions with two parameters a and b. Given $a \in [0,1]$ and $b > 1$, the joint density function is defined as

$$f(x_1,x_2) = \begin{cases} a + (1-a)(1+b)^2(x_1x_2)^b & 0 \le x_1 \le 1, 0 \le x_2 \le 1, \\ 0 & \text{otherwise.} \end{cases}$$

The distribution function F corresponding to f is obtained as

$$F(x_1,x_2) = x_1x_2(a + (1-a)x_1^b x_2^b).$$

Since $F(1,1) = 1$ for any $a \in [0,1]$ and $b > 1$, F is a distribution function.

Next we show that random variables X_1 and X_2 distributed according to f are affiliated. It is sufficient to prove that $\frac{\partial^2}{\partial x_1 \partial x_2} log(f(x_1,x_2)) > 0$ for any x_1,$x_2 \in [0,1]$ (for example, see Krishna ([2002]) and Menezes and Monteiro ([2005])). Differentiating $log(f(x_1,x_2))$ at x_1 and x_2, we have

$$\frac{\partial^2}{\partial x_1 \partial x_2} log(f(x_1,x_2)) = \frac{(1-a)b(1+b)^2 x_1 (x_1x_2)^b}{x_2(a + (1-a)(1+b)^2(x_1x_2)^b},$$

which is positive if $0 \le a \le 1$ and $b > 1$.

The correlation coefficient ρ between two signals is calculated as follows:

$$\frac{36a(1-a)b^2(2+b)^2(3+b)^2}{(12 + b(12 + a(4 + b(4(4+b) - 3a(3+b))))^2}$$

In particular, when $a = 0$, it equals to zero and each signal is independently distributed.

We set $b = 10$ in the following. Figure 1 shows the relation between the correlation coefficient ρ and the value of parameter a when $b = 10$. Note that ρ is increasing in a. As we will see below, the expected revenues of cheating and non-cheating sellers depend crucially on the correlation coefficient ρ as well as the value of belief μ.

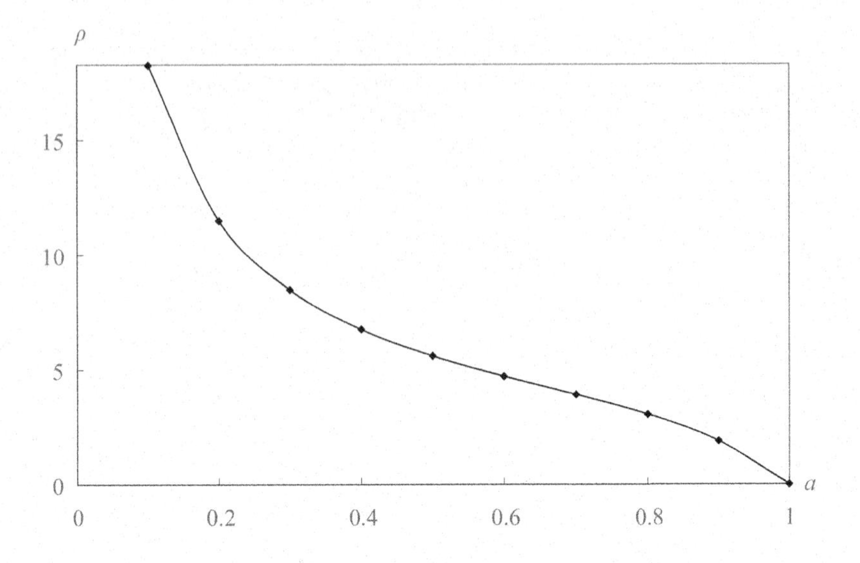

Fig. 1. relation between parameter a and correation ρ when $b = 10$

4.2 Beliefs and Expected Revenues

Figure 2 illustrates how the expected revenues of a cheating seller $E^C(\mu)$ and a non-cheating seller $E^{NC}(\mu)$ change depending on the value of belief μ when $a = 0.4$ and $b = 10$. We also plot the expected revenue from a first price auction E^1 in the figure.

First, both the expected revenue of a cheating seller and that of a non-cheating seller from a second price auction are strictly decreasing as the value of μ increases. This is because the bidders shave their bids more aggressively as they believe that the seller is cheating more confidently (see the equilibrium strategy of each bidder in Proposition 1). Second, the expected revenue of a cheating seller from a second price auction is greater than that from a first price auction, except at $\mu = 1$ in which these two revenues are the same because the equilibrium strategy of a bidder in a second price auction when bidders beleive that the seller always cheats is equivalent to that in a first price auction. Finally, the expected revenue of a non-cheating seller from a second price auction intersects the expected revenue from a first price auction in a certain value of the belief. We denote this value by $\bar{\mu}$ which is equal to 0.21 in this example. If the belief of bidders for seller's cheating is larger than $\bar{\mu}$, the expected revenue from a first price auction is greater than that from a second price auction for a non-cheating seller. In this range of beliefs, as Rothkopf, Teisberg and Kahn ([1990]), Rothkopf and Harstad ([1995]), and Porter and Shoham ([2005]) claimed, a second price auction has the disadvantage to a first price price auction for an honest

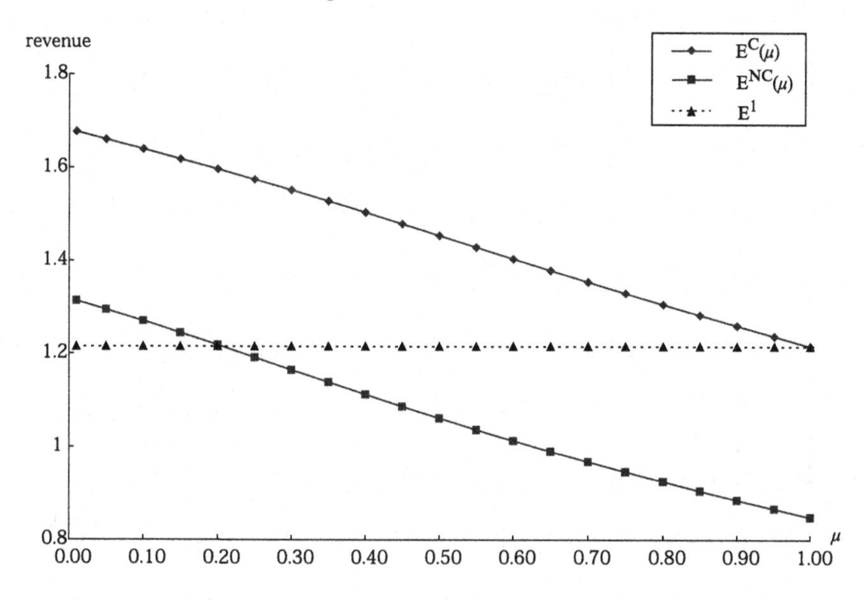

Fig. 2. belief μ and expected revenues $E^C(\mu)$, $E^{NC}(\mu)$ and E^1 when $a = 0.4$ and $b = 10$

seller who does not cheat. However, if the belief of cheating is smaller than $\bar{\mu}$, then the expected revenue from a second price auction is greater than that from a first price auction. Hence, in contrast to the past studies, a second price auction has the advantage to a first price auction if the possibility of cheating predicted by bidders is sufficiently small.

4.3 Correlation Coefficients and Expected Revenues

Figure 3 shows how the expected revenue of a cheating seller $E^C(0.1)$ and a non-cheating seller $E^{NC}(0.1)$ from a second price auction and the expected revenue from a first price auction E^1 change in the value of correlation coefficient ρ when $\mu = 0.1$ and $\mu = 0$. We also plot the expected revenues of a non-cheating seller from second price auction without the possibility of cheating $E^{NC}(0)$ in the figure.

First, all of the expected revenues, that of a cheating seller and a non-cheating seller from a second price auction and that from a first price auction, are strictly increasing in the correlation coefficient ρ because of the effect of affiliation of values. Second, we observe that the expected revenue of a cheating seller is strictly greater than that from a first price auction for any correlation coefficient ρ. This means that the cheating seller always prefers a second price auction to a first price auction because of the cheating behavior. Third, the expected revenue from a second price auction without the possibility of cheating $E^{NC}(0)$ is equivalent to that from a first price auction E^1 at $\rho = 0$ because of the revenue equivalent theorem, that is the expected revenues from both a first price and a second price auction are equivalent when the values of bidders are independent. Moreover, we find that the expected

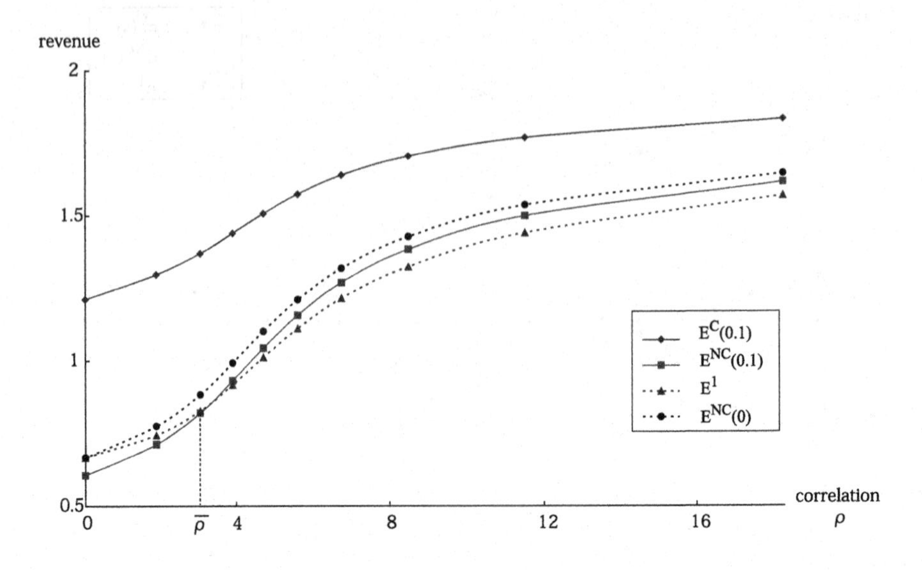

Fig. 3. expected revenues and correlation coefficient of signals for bidders ρ

revenue from a second price auction is always greater than without the possibility of cheating is strictly greater than that from a first price auction when $\rho > 0$. This is because the increment of the expected revenue from a second price auction without the possibility of cheating is greater than that from a first price auction as increasing ρ. Finally, the expected revenues of a non-cheating seller from a second price auction is less than that from a first price auction when $\rho < \bar{\rho}$ but greater than that when $\rho > \bar{\rho}$. When the values of bidders are independent ($\rho = 0$), the expected revenues of a non-cheating seller from a second price auction is smaller than that from a first price auction. If the correlation coefficient is small, a second price auction has the disadvantage to a first price price auction for a non-cheating-seller. However, if the correlation coefficient ρ is sufficiently large, it is not true.

5 Concluding Remarks

In this paper, we compare expected revenues of a non-cheating seller to a first price when the values of bidders are affiliated and the bidders predict possibility of cheating and shave their bids. By a numerical example we show that the expected revenues of a non-cheating seller in a second price auction is still larger than a first price auction if beliefs of bidders for cheating is small and correlation coefficient of signals is not small.

Second price auctions indeed *were* seldom used when Rothkopf, Teisberg and Kahn [1990] has investigated the first analysis of cheating for sellers. However, network technology and inter-net environment has been changing auctions in the past decade. Automatic bidding system emerged in many inter-net auctions can be regarded as a second price auction (see Ockenfels and Roth ([2005])) and it is used in many places. Thus, second price auctions is not rare now. But we should notice that one of important contributions of these studies is to point out fear of cheating and to assert necessity for preventing auctions from cheating. Before 1990's, we do not draw attention to cheating so much, but possibility of cheating in the Internet auctions becomes the central issues in information sciences and electric commerce. Numerous attempts have been made to prevent the cheating for the Internet auction, especially in the field of network security and cryptographic technology. (for example Kikuchi, Harkavy and Tygar ([1982]), Subramainian ([1998]), Liaw and Lin ([2005]).) Even in the field of game theory, some studies about cheating was done, for example, Yokoo Sakurai and Matsubara ([2001]) investigated the false name bids in multi-object auctions. Thus, it is necessary to develop the studies for preventing auctions from cheating by cooperation of many fields.

In conclusion, we want to assert that a second price auction still has the advantage of the expected revenue to a first price auction, especially when the possibility of cheating is sufficiently small and the correlation coefficient of signals is sufficiently large. Some previous studies pointed out that a second price auction is rarely used and that one of the reasons is its possibility of cheating, but its rareness has been diminishing in the past decade. We show that revenue of a second price auction is still larger than that of a first price, even if its cheating effect is considered.

References

[1982] Kikuchi, H., Harkavy, M. and Tygar, J. D. (1988) "*Multi-round anonymous auction,*" in: Proceedings of the First IEEE Workshop on Dependable and Real-Time Ecommerce Systems, 62-69.

[2002] Krishna, V. (2002). *Auction Theory*, Academic Press.

[2005] Liaw, H., Juang, W. and Lin, C. (2005). "*An Electronic Online Bidding Auction Protocol with Both Security and Efficiency,*" Applied Mathematics and Computation, forthcoming.

[2005] Menezes, F. M. and Monteiro, P. K. (2005). *An Introduction to Auction Theory*, Oxford University Press.

[1982] Milgrom, P. and Weber, R. (1982). "*A Theory of Auctions and Competitive Bidding,*" Econometrica, **50**, 1089-1122.

[2005] Porter, R. and Shoham, Y. (2005). "*On Cheating in Sealed-Bid Auctions,*" Journal of Decision Support Systems, **10**, 41-54.

[2005] Ockenfels, A. and Roth, A. E. (2005)." *Late and multiple bidding in second price Internet auctions: Theory and evidence concerning different rules for ending an auction,*" forthcoming to Games and Economic Behavior.

[1990] Rothkopf, M. H., Teisberg, T. J., and Kahn, E. P. (1990). "*Why Are Vickrey Auctions Rare?*" Journal of Political Economy, **98**, 94-109.

[1995] Rothkopf, M. H. and Harstad, R. M. (1995). "*Two Models of Bid-Taker Cheating in Vickrey Auctions*" Journal of Business, **68**, 257-267.

[1998] Subramainian, S. (1998). "*Design and Verification of a Secure electronic Auction Protocol,*" in : IEEE 17th Symposium on Reliable Distributed Systems, 204-210.

[1961] Vickrey, W. (1961). "*Counterspeculation, Auctions, and Competitive Sealed Tenders,*" Journal of Finance, **16**, 8-37.

[2001] Yokoo, M., Sakurai, Y. and Matsubara, S. (2001). "*Robust Combinatorial Auction Protocol against False-Name Bids,*" Artificial Intelligence, 167-181.

A framework for comparing e-business and e-government website initiatives: An empirical study in Spain

Pedro Soto-Acosta and Angel Merono-Cerdan

Dpto. de Organizacion de Empresas y Finanzas,
Universidad de Murcia,
Campus de Espinardo, 30.100 Murcia, Spain.
psoto, angelmer@um.es

Summary. The advent of the Internet and the explosion of e-business in the private sector have brought upon public organizations increasing pressure and incentive to embrace the technology as a means to improve the way governments operate. Since the late-1990s, governments and businesses are shifting to a digital basis for doing business. E-Government programs are diverse and are difficult to assess and compare with e-Business initiatives. There is therefore a need for some framework to allow for assessment and comparison of e-government and e-business initiatives. This paper develops a framework (applicable to both business and governments) that allows evaluation and comparison of the development of the external, public side of their websites. This framework has been applied to study differences in public side of websites between SMEs and local governments in Spain. To achieve this objective, a sample comprising 33 councils and 180 SMEs from the Region of Murcia, Spain was employed. Broadly, the results show that local governments have richer and more advance websites than SMEs.

1 Introduction

The Internet has emerged as a key channel for both firms and governments. Electronic government (e-government) refers to the use of information and communication technologies, and particularly the Internet, as a tool to achieve better government [14]. The impact of e-government at the broadest level is simply "better government" by enabling better policy outcomes, higher quality services, and greater engagement with citizens and businesses [14]. Government organizations at all levels are currently developing and implementing e-government initiatives. In this way, municipal governments are facing the pressure from their clients (i.e., citizens and businesses), partners, and higher-level governmental organizations. It seems that local governments cannot escape this e-government trend. Although e-government can provide communications, transactions, and integration of administrative services, many countries are not making extensive use of the web. A study by Accenture [1]

P. Soto-Acosta and A. Merono-Cerdan: *A framework for comparing e-business and e-government website initiatives: An empirical study in Spain*, Studies in Computational Intelligence (SCI) **110**, 73–86 (2008)
www.springerlink.com

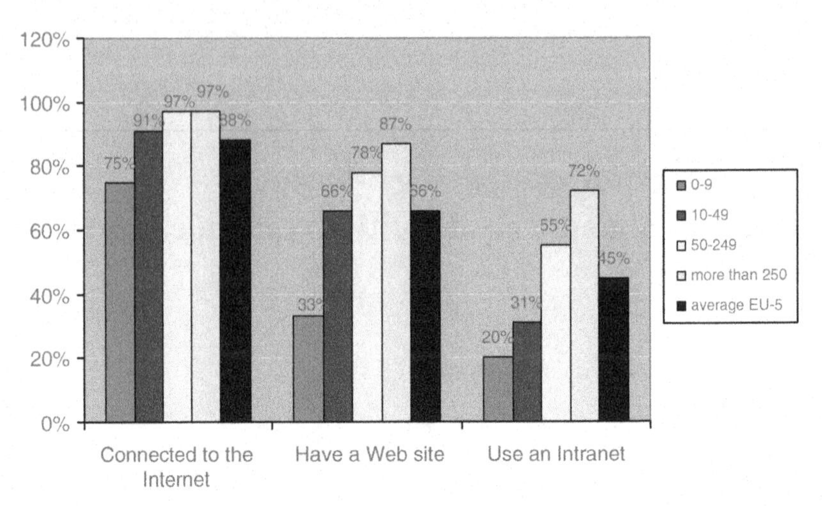

Fig. 1. Access and use of the Internet in European firms by number of employees (Source: e-Business W@tch, 2004)

found that the world's governments are at a crossroads with their online programs. With a few exceptions, e-government advances have slowed over the past few years [1]. While some e-government strategic agendas focus primarily on information and service delivery issues, other may focus more on creating internally efficient systems and processes. Still other may adopt a more comprehensive view, incorporating issues such as constituent relationship management and e-democracy [4]. Although each of these views of e-government may be legitimate, there is a need for some common understanding to allow for assessment, comparison, and explanation of current and future efforts in e-government initiatives. In this sense, we find particularly useful to take as reference the progress achieved in the analysis of Internet implementation within the private sector.

E-business can be defined as the management of relationships, electronic data interchange, collaboration, communication and the establishment of workflow processes with business partners, customers, employees, government and other business agents, as long as these tasks or processes are performed by electronic means. This definition is broadly consistent with e-business definitions found in the literature [9][16][17][21]. In this way, e-business represents the conduct of business activities over the internet. Today, organizations' Web presence is no longer exclusive to large companies or highly innovative firms. This statement is supported by the high rates of Internet adoption among firms. For example, Figure 1 shows 88% of European companies in EU-5 (Germany, Spain, France, Italy, and the UK) are connected to the Internet with no difference regarding business size when considering companies with more than nine employees.

The literature on Web adoption recognizes the adoption of an e-mail account as the minimum Web adoption level [18][19]. These firms are normally connected to the

AQ: Please check running head.

Internet and have an e-mail account that they use to establish links with customers and business partners. Nonetheless, creating a Web site is the starting point for a firm to achieve the benefits derived from using the Internet. The average of firms with Web site in EU-5 is two thirds, with a clear relationship between having a Web presence and business size (considered as number of employees) (see Figure 1). As a result, only one third of companies with fewer than 10 employees have a Web site, while this number is doubled when considering businesses with more than ten employees. The next, more advanced, level in the use of the Internet is the establishment of a corporate intranet. Figure 1 shows that 45% of companies in EU-5 have an intranet, with 50 employees the inflexion point at which businesses with an intranet are in the majority.

In recent years, the research areas of e-government and e-business have experienced an exploding growth because of its relative novelty, global reach, and economic impact. Although using a common background concerning research methods, the two are traditionally studied separately. In this sense, recent research has started to suggest that e-business and e-government are related and should be studied together.

In an attempt to cover this research gap, this study develops a methodology (applicable to both private corporations and governments) that allows evaluating and comparing the development of the external public side of websites. That is, the main objective is directed to the valuation of the public website sections, those which are publicly accessible, according to the design and content of the website. In this regard, content characteristics are analyzed under three web orientations (informative, relational, and transactional).

To compare e-government and e-business web development, the organizations selected for the study are local governments and SMEs (small and medium size enterprises). Local governments and SMEs share that both have less financial, technological and personnel resources than their higher-level counterparts (large firms and national or regional governments).

The nation considered for this research is Spain. A study by e-business w@tch [3] suggested that companies from this country have similar infrastructure concerning information and communications technologies (ICTs) than other European Union (EU) member states such as France and Italy. This report also indicated that the geographic divide in e-business activity, within the EU, was smaller than initially expected. With Spanish firms having close e-business figures in comparison with France and higher than Italian companies. With regard to e-government statistics, Spain is well-positioned internationally with similar online availability of public services to France and higher than Germany and Italy [14].

Considering the above-mentioned points, the key research questions that motivated our work are:

- Is it possible to draw up a framework capable of evaluating and comparing e-government and e-business website initiatives?
- What is the real use of websites by SMEs and local governments in Spain?

- What are the differences, if any, between e-government and e-business websites initiatives in Spain?

This paper consists of six sections and is structured as follows. The next section presents the study's theoretical foundation. Then, the methodology used for the sample selection and the data collection are discussed. Following this, the data analysis and the empirical results are examined. Finally, the paper concludes with a discussion of research findings, limitations, and directions for future research.

2 Literature review

A great deal of work has been devoted to investigations of the use of Web sites for commercial purposes [6][11]. Web sites can be used for other reasons besides e-commerce. They can provide company information and offer the necessary means to communicate with the company [22]. In this respect, Heldal et al. [5] developed a model that shows how communications can be optimized through the corporate site by improving and developing sustainable relationships. Therefore, it can be argued that corporate Web sites frequently represent the firm's strategic intent to use the Internet (especially when considering SMEs) in order to share information, facilitate transactions, communicate with different stakeholders and improve customer service.

The minimum externally accessible content in any given website, either from the public or private sector, is related to the offering of basic contact information such as name, address, electronic mail and phone number. In this regard, Teo and Pian [18] found the presence of contact methods on businesses' websites do not differ much among the web adoption levels. Nonetheless, when engaging in further website development these entities (firms or government agencies) usually present content features used for other purposes such as to facilitate background information, communicate and establish relationships (with customers, citizens, businesses, the general public etc.), provide information on products and/or services, or even present transactional features for the fulfillment of orders or government requirements online. Given the different nature of these content characteristics, website development presents distinct directions for benchmarking and improvement.

Within the literature on web development, the primary orientations that have been studied for web content analysis are: information, communication and transactional orientation. In Table 2, a group of studies related to the topic of this paper have been structured according to the type of analysis used. Thus, the content categories considered in each research for the website analysis are identified and classified according to the mentioned web orientations. Huzingh [7] conducted empirical research of a sample made up of companies found on Yahoo and the Dutch Yellow Pages. He considered information, transaction and entertainment as the main features of content. Robbins and Stylianou's [15] work is an extension of Huzingh's and was intended to overcome many of its weaknesses. The sampling frame was designed to be a good mix of international companies and a more comprehensive set of content features

Web orientations	[5]	[11]	[13]	[17]
Informative	Information	Informational	Corporate & Commercial information	Information (products)
Relational		Communicational	Communication / Customer support	Interaction
Transactional	Transaction	Transactional		Transaction
Other	Entertainment			Supplier Connection

Fig. 2. Content features considered in previous research

was employed. Other authors, such as Zhu and Kraemer [23] used a web content analysis to characterize a firm's e-commerce capabilities. E-commerce capabilities were measured along four dimensions: information on products and services, transaction, interaction and customisation, and supplier connection. They thereby created a total e-commerce capability index by aggregating these four composite metrics. In addition, attempts to evaluate the content quality of a website have also been made. Miranda and Banegil [13] considered three sets of factors to assess the content quality of a website: informational, communicational, and transactional. They thus created a web assessment Index which focused on four categories (accessibility, speed, navigability, and content).

3 Research framework

To evaluate and compare SMEs and local government websites in the following sections a research framework which assesses content and design characteristics is introduced. For this research, content and design features were measured objectively (e.g. the website either contained the feature or did not). Table 3 provides an overview of the features measured.

3.1 Content features

This study intends to measure website content in both firms and governments according to the mentioned web orientations. With this objective in mind, we next present our approach for each of these web orientations.

Informative orientation

Firms can use their websites to disseminate corporate or commercial information to customers, business partners or other stakeholders (shareholders, employees, the

Content features	Design and privacy features
Contact methods	**Navigation**
Address	Well-structured menus
E-mail	Site map
Phone number	Search function
	Use of multimedia
Informative orientation	**Privacy**
-Corporate/institucional information	Privacy statement
History	
Message from the CEO/mayor	
Organizacional charts	
Financial report	
Employment opportunities	
-Commercial/service information	
Product/service description	
Product/service prices	
Relational orientation	
Client support (e-mail or form)	
Webmaster (e-mail or form)	
Internet chat	
web forum	
Registration for newsletter	
Transactional orientation	
Online ordering	
Ordering by E-mail or phone	
Ordering by Electronic form	
Electronic payment	

Fig. 3. Website content and design features

public, etc.) [7]. Corporate information can provide insight into the background of the company (financial statements, employment offers, history, etc.) and commercial information implies providing product-related information, such as prices, specifications, terms of delivery, etc. Similarly, governments use their websites to diffuse information on the web. In fact, as many citizens and business are able to access information from the private sector, they expect the same access from the government. Government information includes two categories of information "institutional and services information". Institutional information refers to the providing of generic data about some general facets of government related to administrative, political, and socioeconomic aspects (government structure, organizational chart, employment, history, etc.). Service information comprises the publishing of information about the services and functions that governments deliver to citizens, businesses, and other stakeholders. By these means, government intends to facilitate detailed information on their services as well as where to go for government services and post-service support. Therefore, the informative orientation consists of one-way electronic information directed to one or more stakeholders. This way, the quality of information on a website, either from businesses or governments, is reflected by the extent of available information on each of theses categories.

Relational orientation

Internet communications besides allowing a cost reduction in comparison to traditional communication tools, they offer a unique and integrated opportunity for interacting with several business agents (both internal and external to the organization). Certain applications such as the Internet chat or the web cam enable a two-way real time information exchange between an organization and its stakeholders (customers, citizens, suppliers, businesses, etc.). Unlike other cost-effective and user-friendly applications such as the e-mail, the web forum, and the feedback form allow unsynchronized two-way conversations. Moreover, the creation of web forums could form the basis of online communities where people can exchange views. In this sense, all these Internet technologies facilitate the exchange of information, collaboration and the possibility of establishing close relationships based on trust and mutual commitment. The difference between the informative orientation and the relational orientation is that the latter permits two-way information exchange. This exchange of information can vary from more structured tools such as the request for information form to more open and interactive forms such as the online chat. As a result, the quality of communication on a website is estimated by the extent of available communication mechanisms through which an agent can interact with the firm/government or with other agents (using its website as the web platform). Presumably, both firms and governments present similar zeal in providing support to customers/citizens and to allow them the opportunity to discuss about products and/or services and public issues or new policies.

Transactional orientation

Over the past 20 years the economy has rapidly transformed from its traditional base to a new, information-based economy. In this new environment, work has shifted from the creation of tangible goods to the flow of information through the value chain [2]. For this transition, the establishing and development of workflow processes has played a fundamental role. According to the Workflow Management Coalition [20] a workflow is "the automation of a business process, in whole or in part, during which documents, information or tasks are passed from one participant to another for action, according to a set of procedural rules". Internet technology provides great opportunity for automation processes. Thus, the transactional orientation is considered a web orientation which involves the establishment of electronic processes for the fulfillment of orders or government services through the firm's website. In ideal cases, web transactions are connected directly to the internally functioning government systems with minimal interaction with government staff [10]. The same argument holds for businesses.

3.2 Design and privacy features

Design refers to the way in which the content is presented to the Internet users. Huzingh [7] considered six characteristics in web design: navigation structure, search function, protected content, quality of the structure, image, and presentation style. Liu and Arnett [11] proposed a framework which considered information quality, learning capability, playfulness, system quality, system use, and service quality as factors that are related to well-designed websites. They tested these factors on a sample of companies from Fortune 1000 and found that information quality, system use, playfulness, and system design influenced the success. Based on prior research, we identified four variables to capture the design dimension of a website: Well-structured menus, site map, internal search engine, and use of multimedia.

As websites use a number of mechanisms to gather information about their visitors, privacy is becoming an important issue. Some of these mechanisms such as registration and ordering forms are explicit. However, other implicit mechanisms for tracking online activities exist (the length of time spent on each page, frequently accessed products, etc.). While this tracking information can be used toward improving the website and its offerings, the temptation to sell this information to external parties may also exist [10]. Thus, organizations must inform on their website about the use of personal information collected from their visitors. For this study, the presence of a privacy statement on the organization's website was examined.

4 Methodology

The authors directly observed the websites of all councils in one Spanish region, the Region of Murcia (Southeast Spain), and a listing of SMEs from the same region. To explore the content and design of websites a content analysis on the organization's (company or municipality) website was performed. Content analysis has been previously applied on few empirically based investigations relating to the e-business topic in the literature [7][15][18][23]. The main contribution of this technique comes from the possibility of measuring objectively a significant number of content features [7].

4.1 Sample

This study employed a sample consisting of 33 councils and 180 SMEs with website from the Region of Murcia (Spain). The diffusion of websites for the councils of this region is nearly 87% (33 out of 38 of its councils have a website). Thus, all the local government websites from this region were analyzed. The selected businesses were firms compelled to present financial statements and had between 1 and 250 employees. Further, organizations pertaining to certain activities within the service industry (NACE groups 80, 85, 90-93) as well as agricultural and fishing companies (NACE groups 1-9) were excluded as these firms do not conform to the habitual concept of business. Taking into account that in the Region of Murcia exist 33,753 companies with at least one employee (excluding those from the above mentioned NACE groups) and considering that several sources [8] predict that in the Murcia region only 28.9% of the companies with more than ten employees have a corporate website, p = 30 was applied. Finally, the sample error estimated was 6.8% at the 95.5% confidence interval.

4.2 Data Collection

To analyze each company's website an electronic questionnaire was developed. This way the data introduction process was intended to be facilitated making it easier, faster and more accurate (required fields were used). The questionnaire evaluated different content and design characteristics. The items were introduced considering previous studies within the literature [7][12][15][18]. A panel of academic experts was also consulted to ensure all variables measured broadly the website's content. The questionnaire was initially pretested on 10 companies and 7 local governments. This process resulted in 24 variables that were used to measure a firm's website content and design.

Each variable represented a different feature and was coded using a binary variable, where 1 was "yes" and 0 was "no". Data was gathered and independently coded by both the authors during April and May 2004. Subsequently, comparisons were made to check for consistencies in coding.

5 Results: content

5.1 Results for Contact methods

The contact method which is most often available is the organization's phone number. Of all analyzed websites, 91.5% contain it (see Table 4). When both sources, business and councils, are compared significant statistical differences were found for the three contact methods. Thus, more businesses than councils presented address, e-mail, and phone number.

5.2 Results for the informative orientation

With regard to corporate/institutional information, councils presented more features than business for all the items (see Table 4). Statistical differences at the 5% level were found for the History item, while significant differences at the 1% level were encountered for message from the Mayor, organizational charts, and employment opportunities. No significant differences were found for the presence of financial statements. The type of corporate/institutional information more frequently presented in

Content features	SMEs		Councils		Chi-squared test		Total	
	N	*%*	*N*	*%*	*Chi-squared*	*p*	*N*	*%*
Contact methods								
1. Address	170	94.4	20	60.6	33.1	0.001**	190	89.2
2. E-mail	150	83.3	22	66.6	4.9	0.032*	177	83.0
3. Phone number	168	93.3	27	81.8	4.8	0.041*	195	91.5
Informative orientation								
-Corporate/institucional information								
4. History	126	70.0	29	87.8	4.5	0.035*	155	72.7
5. Message from the CEO/mayor	17	9.4	20	60.6	50.8	0.000**	37	17.3
6. Organizacional charts	29	16.1	23	69.6	43.3	0.000**	52	24.4
7. Financial report	12	6.6	5	15.1	2.7	0.152	17	7.9
8. Employment opportunities	26	14.4	16	48.4	20.4	0.000**	42	19.7
-Commercial/service information								
9. Product/service description	121	67.2	26	78.7	1.7	0.223	147	69.0
10. Product/service prices	42	23.3	5	15.1	1.1	0.366	47	22.0
-Communicative orientation								
11. Client support (e-mail or form)	125	69.4	19	57.5	1.7	0.225	144	67.6
12. Webmaster (e-mail or form)	48	26.6	13	39.4	2.2	0.147	61	28.6
13. Internet chat	2	1.1	5	15.1	17.3	0.000**	7	3.2
14. web forum	3	1.6	7	21.2	23.8	0.000**	10	4.7
15. Registration for newsletter	12	6.6	6	18.1	4.7	0.041*	18	8.4
-Transactional orientation								
16. Online ordering	37	20.5	25	75.7	41.1	0.000**	62	29.1
17. Ordering by E-mail or phone	28	15.5	24	72.7	49.3	0.000**	52	24.4
18. Ordering by Electronic form	23	12.7	5	15.1	0.1	0.779	28	13.1
19. Electronic payment	13	7.2	0	0.0	2.5	0.227	13	6.1

*Significant at $P < 0.05$ level.
**Signigicant at $P < 0.01$ level.

Fig. 4. Content features of the website versus the source of the site

both types of websites is by far History with 72.7% of the total number of websites containing this feature. As far as commercial/service information is concerned no significant differences between councils and businesses were found for the two variables. When considering both groups simultaneously more than two-third of websites contained a detailed product/service description, while slightly less than one fourth presented their product/services prices.

5.3 Results for the relational orientation

The Relational orientation has been measured by five variables: Client support, Webmaster e-mail/form, Internet chat, web forum, and registration for newsletters. As shown in Table 4, more than two-third of the websites were found to contain an e-mail or feedback form specific to client support. Less than 30% of websites offered the possibility of contacting the webmaster, while the other three characteristics (Internet chat, web forum, and registration for newsletters) were found in less than 10% of the websites. However no statistical differences (between businesses and councils) were found for the Client support and Webmaster features, significant differences (all in favour of councils) were found for the rest of the relational characteristics.

5.4 Results for the transactional orientation

These content characteristics are related to the possibility of fulfilling orders or government requirements online. As presented in Table 4, nearly 30% of the websites permitted online ordering. Nonetheless, only 7.2% of the business websites allowed electronic payment, while no council website had this characteristic. When both sources, business and councils, are compared significant statistical differences where found for online ordering and ordering by e-mail or phone. Thus, more councils than businesses presented the chance of conducting government requirements and services online, but these services were offered primarily by e-mail or telephone.

6 Results: design and privacy

Of the variables related to the design of a website, it was found that over 90% of the websites had well-structured navigation menus (see Table 5) and statistical differences for this characteristic were not found between the two sources (SMEs and councils). Statistical differences at the 1% level were found for the other three variables: site map, search function, and use of multimedia. In this way, more council's websites contain site map and search function, while more SMEs include multimedia features on their websites. Regarding privacy issues, only 22% of the total number of websites analyzed contain a privacy statement. Nonetheless, no statistical differences for this feature were found between SMEs and councils.

7 Discussion and Conclusions

Although much research has been conducted into different e-business and e-government issues, there is a need to further investigate into more basic and primary use of the Internet, the external website as a means of interaction with stakeholders. Moreover, despite using a common background concerning research methods, e-business and e-government have traditionally studied separately. To cover this research gap, this paper develops a framework that allows evaluation and comparison of external web development of business and government websites. The organizations selected for this study are local governments and SMEs because of their closeness and level of expertise concerning Internet implementation. Broadly, this research offers several contributions:

Design and privacy features	SMEs		Councils		Chi-squared test		Total	
	N	*%*	*N*	*%*	*Chi-squared*	*p*	*N*	*%*
Navigation								
20. Well-structured menus	167	92.7	27	81.8	4.1	0.088	194	91.0
21. Site map	41	22.7	16	48.4	9.4	0.005**	57	26.7
22. Search function	27	15.0	15	45.4	16.3	0.000**	42	19.7
23. Use of multimedia	100	55.5	9	27.2	8.9	0.004**	109	51.1
Privacy								
24. Privacy statement	42	23.3	5	15.1	1.0	0.366	47	22.0

*Significant at $P < 0.05$ level.
**Signigicant at $P < 0.01$ level.

Fig. 5. Design features of the website versus the source of the site

1. It facilitates a framework for evaluating and comparing e-government and e-business websites initiatives;
2. It provides sufficient knowledge about the real use of websites by SMEs and local governments in Spain;
3. It sheds light on which are the main differences between SMEs and local governments in Spain.

The results show that SMEs' websites contain more contact methods (address, e-mail, and phone number) than councils. This could be interpreted such that businesses are more willing to be sure that potential customers have different means to get in touch with them. Unlike councils seem to prefer being contacted by certain contact methods (mainly by phone). This finding contrast with Teo and Pian's [18], which suggested that within firms contact methods do not differ among the web adoption levels. Therefore, Teo and Pian's [18] finding does not seem to hold when considering websites pertaining to organizations with different goals.

The empirical results also demonstrate that councils present on their websites more corporate/institutional information than SMEs. Significant differences were found for four out of five websites features of this category. This implies that councils are more interested in disseminating information about themselves, while businesses may be more concerned with providing information on their products and services. The reasons may be because, first, although this type of information does not include real services for the citizens, it can be useful for other purposes of councils such as attracting visitors. Second, governments are more committed with information transparency and social responsibility. With respect to commercial/service information, significant differences between SMEs and councils were not found. Thus, presumably councils and SMEs have similar zeal in providing information about their products/services.

Although within the relational web orientation differences for client support and webmaster e-mail were not found, differences were encountered for the other three variables: the presence of Internet chat, web forum, and registration for newsletter.

This argument together with the results obtained for the design features (where was found that more council's websites contain site map and search function) indicate that councils have richer and more technologically advanced websites. Thus, councils engage further than SMEs in web development by allocating more funds and hiring top web design companies. In addition, with regard to the transactional web orientation, this study found that more council's websites allowed the possibility of fulfilling services or government requirements online. Trust can be an important issue here. In this sense, governments are more trustworthy organizations than businesses when considering online payment and services.

While the study's contributions are significant, it has some obvious limitations which can be addressed in future research. First, websites have been analyzed at one point in time (cross-sectional picture) while the web is a highly dynamic medium. Therefore, a longitudinal study could enrich the findings. Second, the country from which our sample was obtained is Spain. Similar studies at different countries are likely to show different results, especially when considering high e-business and e-government intensity countries such as the USA, Finland, and Canada. Therefore, in future research, using a sampling frame which combines councils and SMEs from different countries could provide a more international perspective to the subject.

References

1. Accenture, Fifth annual accenture eGovernment study (2004).
2. Basu, A., Kumar, A.: Research commentary: workflow management issues in e-business. Information Systems Research, Vol. 13 1 (2002) 1-14.
3. E-business W@tch: The European E-business report (2004).
4. Grant, G., Chau, D.: Developing a generic framework for E-government. Journal of Global Information Management, Vol. 13 1 (2005) 1-30.
5. Heldal, F., Sjovold, E., Heldal, A.F.: Success on the Internet: Optimizing Relationships through the Corporate Site. International Journal of Information Management, Vol. 24 (2004) 115-129.
6. Hoffman, D.L., Novak, T.P., Chatterjee, P. et al.: 1997: Commercial scenarios for the Web: opportunities and challenges. Journal of Computer-Mediated Communication, Vol. 3 1 (1997), 1-21.
7. Huzingh, E.: The Content and Design of Websites: An Empirical Study. Information & Management, Vol. 37 3 (2000) 123-134.
8. INE: Encuesta de uso de Tecnologias de Informacion y Comercio Electronico (2002) (http://www.ine.es).
9. Kalakota, R., Robinson, M: E-business roadmap for success. Addison Wesley Longman (2000).
10. Layne, K., Lee, J.: Developing fully functional E-government: A four stage model. Government Information Quarterly, Vol. 18 (2001) 122-136.
11. Liu, C., Arnett, K.P., Capella, L.M., Beatty, R.C.: Websites of the Fortune 500 Companies: Facing Customers through Homepages. Information & Management, Vol. 31 6 (1997) 335-345.
12. Merono, A.L., Sabater, R.: Valoracion del nivel de negocio electronico. Revista Europea de Direccion y Economia de la Empresa, Vol. 12 1 (2003) 9-22.

13. Miranda-Gonzalez, F.J., Banegil-Palacios, T.M.: Quantitative Evaluation of Commercial Websites: an Empirical Study of Spanish Firms. International Journal of Information Management, Vol. 24 4 (2004) 313-328.
14. OECD: The e-Government imperative (2003).
15. Robbins S.S., Stylianou, A.C. (2003): Global Corporate Websites: An Empirical Investigation of Content and Design. Information & Management, Vol. 40 3 (2003) 205-212.
16. Rodgers, J.A., Yen, D.C., Chou, D.C: Developing e-business: a strategic approach. Information Management and Computer Security, Vol. 10 4 (2002) 184-192.
17. Sawhney, M., Zabin, J.: The Seven Steps to Nirvana, New York: Free Press (2001).
18. Teo, T.S.H., Pian, Y.: A Model for Web Adoption. Information & Management, Vol. 41 4 (2004) 457-468.
19. Teo, T.S.H., Tan, M., Wong, K B: A contingency model of Internet adoption in Singapore. International Journal of Electronic commerce, Vol. 2 2 (1998) 5-18.
20. WFMC - Workflow Management Coalition- (2004) (http://wfmc.org).
21. Wu, F., Mahajan, V. and Balasubamanian, S.: An analysis of e-business adoption and its impacts on business performance. Journal of the Academy of Marketing Science, Vol. 31 4 2003, 425-447.
22. Young, D., Benamati, J.: Differences in Public Websites: The Current State of Large U.S. Firm. Journal of Electronic Commerce Research, Vol.1 3 (2000) 94-105.
23. Zhu, K., Kraemer, K.L. E-commerce Metrics for Net-enhanced Organizations: Assessing the Value of E-commerce to Firm Performance in the Manufacturing Sector. Information Systems Research, Vol. 13 3 (2002) 275-295.

A Transactional Relationship Visualization System In Internet Auctions

Masao Kobayashi[1] and Takayuki Ito[2]

[1] Department of Computer Science,
Nagoya Institute of Technology,
Gokiso, Showaku, Nagoya 466-8555, JAPAN.
kobayashi@longwood.mta.nitech.ac.jp

[2] Techno-Business School,
Nagoya Institute of Technology,
Gokiso, Showaku, Nagoya 466-8555, JAPAN.
ito.takayuki@nitech.ac.jp

Summary. As Internet auctions have increased, so too has auction fraud. This paper describes the design of a system that supports fraud detection and market prediction by visualizing transaction networks on Internet auctions. Our system employs link mining techniques and user information. At this stage, we successfully showed visualized trading networks by extracting trading histories from an auction site. Using a visualized graph, the system shows suspiciousness with user ID information. Also, the system presents trading relationships as a network structure in various viewpoints. Furthermore, the system possesses an explanation function on a visualized trading network that predicts which buyers and sellers are active and did better behaviors. Our preliminary experiment demonstrates that the graph presentation function is scalable enough against the number of sellers and buyers.

1 Introduction

In recent years, as Internet auctions increase, reports of such illegal acts as fraud have also begun to surface in the media. In its monthly report for March 2007, Yahoo! Auction [5], the largest Japanese Internet auction site, announced that the number of monthly login IDs has surpassed 19 million and the number of exhibitions at auctions is over 13 million cases per day. Furthermore, the number of login IDs exceeds 5 million in only an auction section. The second largest auction site in Japan, Rakuten Auction [6], has 3 million IDs. Unfortunately, fraud victims are also increasing. In 2006 December, over 990 cases of fraud were reported totaling losses of 88 million JPY (about $725,100). The trick of fraud used by this case was generally well known. This trick is that as for the early period of use of auction, the frauder have dealings commonly, but commit fraud when they get trust. Yahoo also identifies illegal acts with user IDs, but such prevention is insufficient.

M. Kobayashi and T. Ito: *A Transactional Relationship Visualization System In Internet Auctions*, Studies in Computational Intelligence (SCI) **110**, 87–99 (2008)
www.springerlink.com

Notice the difficulty of judging how to sell or bid safely and make a better market because in Internet auctions, relations between persons or between a person and an article are immediate. Therefore, we decided to find new values by treating an Internet auction as a single network. We intended to find a similarity with an existing problem by regarding Internet auction as a network structure.

In addition to network structure, we focus on user reputation information provided by other users. Most auction and shopping sites are equipped such reputation mechanisms. Such ratings unfortunately lack exact and correct information. However, we can learn how the player acts based on his/her trading history. This information warns novices to be sensitive and aware of auction participation. We also utilize other extractable information from auction sites, such as the amount of buying, selling, and bidding.

In this work, the transaction relationship on Internet auction sites is treated as a network structure with information, and we present a user-friendly visualization system that has the following three points:

1. This system supports the identification of fraud from transaction relationships and detects suspicious behavior with such user ID information as selling/buying history.
2. This system presents a transaction relationship as a network structure from various viewpoints. Users can view the network from various angles such as seller or buyer.
3. This system has explanation functions about the visualized network. This function predicts auction markets.

Many researchers have realized the importance of fraud detection and methods and systems. There are some similar works. NetProbe [19] extracts a network from e-bay auction sites and tries to identify possible frauds. They actually only focus on network structure without utilizing such user information as reputation ratings, selling/buying histories, etc. We, however, use such private information to identify suspicious acts. Other related work will be shown in Section 2.

The rest of this paper is organized as follows. We review related work and the present condition of auction fraud in Section 2. Then we describe the detailed system architecture in Section 3. Next, in Section 4, we report experiments that evaluate this system with user IDs on actual auction sites. In this paper, we also evaluated execute time. In Section 5, we discuss future systems and evaluation methods. Finally, we summarize our findings in Section 6.

2 Related Work

In this section, we survey related approaches for fraud detection in auction sites as well as the literature on reputation systems typically used by auction sites to prevent fraud. We also look at fraud on the Internet and related works on trust or authority propagation, reputation systems, and link mining that could be applied to predict network links. In addition, we describe related work on market prediction and theoretical results.

2.1 Fraud on the Internet

In recent years, fraud in Internet auction sites in Japan has been investigated by the Japanese Metropolitan Police Department. It is a fraud in Internet auction sites that is performed with unjust access most [26]. According to the breakdown, Internet auction manipulation accounts for 60% of all illegal access. Fraud in auction sites can be classified into several patterns [15], including spoofing [27] and phishing [28]. In the past, people tried to identify potential fraudulent individuals. However, methods for arresting these criminals use such "classic"or "common sense" tactics as newspapers articles [9] and law enforcement agencies [10]. In newspapers fraud victims demand reporting of cases with newspaper articles. There are also auction sites fighting frauder in those own rights [11]. There is such an effect in continuing approaching these, but needs cost likes time and money very much.

2.2 Trust or authority propagation

Trust propagation is used by TrustRank [1] to detect Web spam. The goal is to distinguish between "good" and "bad" sites. Because web spam pages employ various ways to achieve higher ranking in search engine results, we filter only "good" pages. Authority propagation has been studied in Web searches. PageRank [2] uses hyperlinks as "votes." In this concept, a page linked by many "good" pages must after all be a "good" page. In effect, the importance of pages is propagated over hyperlinks connecting them. HITS [3] represent this concept as hub and authority pages. Authority indicates abundant information about a specific topic. Hub indicates that links to high value pages have authority, which is abundant. In general, "good" sites about a specific topic have a high evaluation value of authority, and a "good" collection of links has a high evaluation value of hub.

2.3 Reputation system

Auction sites use reputation systems to prevent fraud. Comment of a user is one of the persuasive evaluation indexes. But they are usually simple and can be easily foiled because it is difficult to detect honest behavior and show that a user's reputation reflects actual intention [1]. Resnick et al. [13] and Melnik et al. [14] show that reputation systems might not be effective to prevent fraud.

2.4 Link mining

Link mining is a kind of data mining and a focus graph structure that consists of nodes and edges. Pattern mining and graph mining resemble link mining [8]. Many kinds of tasks are included such as link prediction, classifying nodes, ranking nodes, and subgraph discovering [16]. The subgraph discovery is a very important task. It is thought that we can use that we find a subgraph having a similar pattern for the detection of similar preference of a user and detection of a fraud group. A link prediction can be defined as: "when a known part of a network structure is given, I predict an unknown part to a clue in this." This idea can be identified as the presence of a link between two nodes in fraud detection.

2.5 Market prediction

Market predictions are widely studied as well as in the field of computer science. Particularly, a recommendation agent system uses collaborative filtering to make recommendations [17]. The collaborative filtering system supposes user tastes based on the taste information of users taking similar actions as the user. In our system, we provide prediction information based on a trading network.

2.6 Theoretical results

Many theoretical results are related to false-name proof auction design. Such auctions [22][23][25] guarantee that no player benefits from virtual or dummy players who make false bids. However, these results are quite theoretical. In particular, they focus on VCG [20][21][24], the most theoretically desirable combinatorial auction that has not yet been employed in the practical world due to several application problems. On the other hand, in our proposed system, we practically construct a system that can reveal cheats or predict targeted markets.

3 Transactional Relationship Visualization System

3.1 Outline of the system

This system consists of three parts. The first part selects transaction relationship records from Internet auction sites. The second displays graphs consisting of the characteristic elements of the extracted records. The third is an instruction part with information from selected transaction records. We show a conception diagram of the system in Figure 1. On this system, the user queries with an auction ID or a product name. With the query entry, the Java program accesses the Internet auction site and analyzes the HTML sources of the auction site pages. The Java programs select data available for visualizing transaction relationships. Then, user ID information from the auction site is stored in the user database. We can chronologically analyze information by adding acquisition data to the time tags. Finally, a visualization transaction relationship network represents users by graphs. We describe system implementations below.

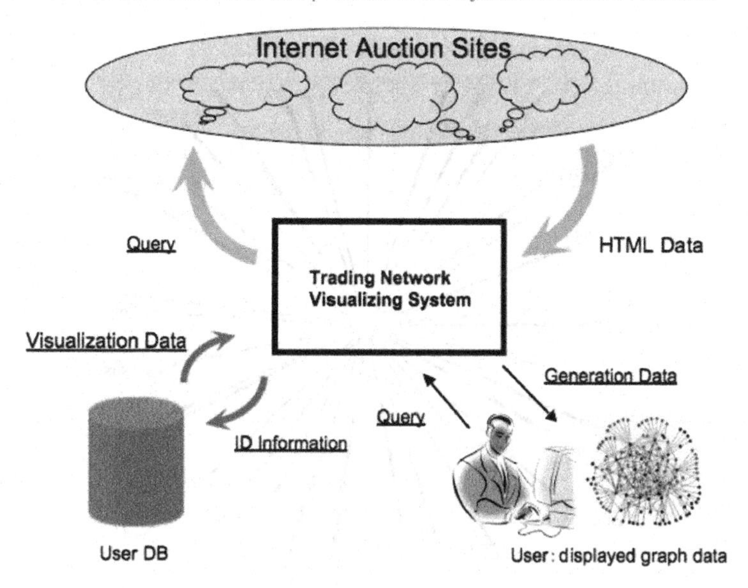

Fig. 1. Outline of proposed system

3.2 Visualization of Transaction Relationship

We first have to get transaction data from Internet auction sites to visualize an auction network. We use Yahoo! Auction, the largest auction site in Japan, for implementation. We obtained transaction data and analyzed the HTML source codes of auction pages to exploit user IDs and articles. This system possesses review pages of personal ratings. If this page describes user A, it represents the ratings of A from partners who dealt with A. Therefore, this page indicates A's transaction history.

Using the Magnetic-spring model [4], we make graphs that consist of nodes and edges. The spring model is one drawing method for a graph. The magnetic-spring model introduces magnetic theory into spring models. Figure 2 represents a graph about certain user IDs written in Java programs.

This part of this system's implementation is written in Java, which uses JRE 1.5.0 as the library. The JLabel class is used to display the nodes. The data structure is Vector class and stores information of nodes and edges. The elasticity of a spring is affected by the distance between two nodes. This panel is clickable, and any node can be freely moved anywhere by drag and drop. Nodes implement MouseListner, and additional operation is available for the graphs.

These edges represent seller and buyer relationships as differences of line width. Here, wide edges indicate that the user ID is a "successful bidder" for the trading partner. Narrow edges indicate that the user sold goods to the trading partner. In addition, double-clicking on the node forms a new graph. That is, search depth can be lowered by clicking the node. In the future we aim to visualize the entire Internet auction network. This function is the first step. Figure 3 indicates a graph with two

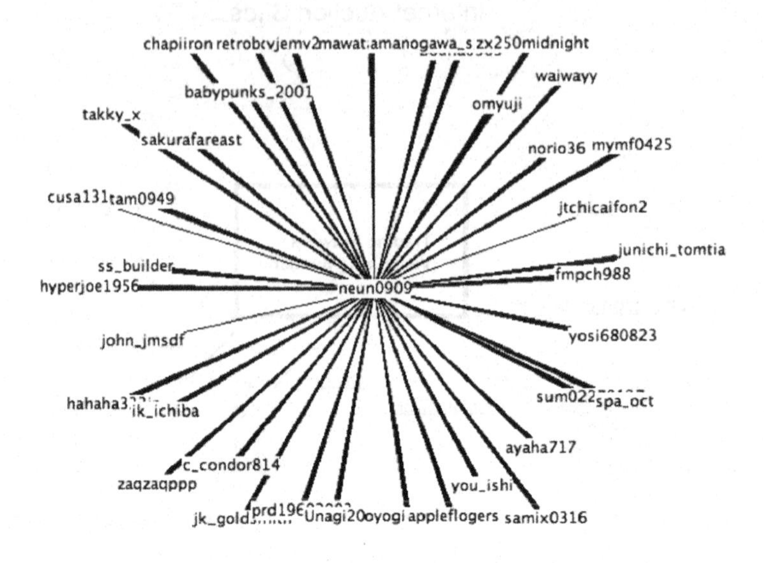

Fig. 2. Visualization of transaction relationship

nodes as search keys. In this figure, the auction ID, "carbanknimura," is rated by 33 transaction partners. In other words, "carbanknimura" has 33 transaction partners. "monkey5551103" is one such partner who is also rated by seven people including "carbanknimura."

User ID rating value is one of the strongest component parts of graphs. Reputation systems are widely used in many Internet auction sites, making rating analysis important. However, the reliability of rating value must be verified in the future.

3.3 Introduction function with transaction information

This function captures the rating value of user IDs and represents the following seven items:

1. rating value of user ID
2. rating value average of traded partners
3. auction type of user ID
4. percentage of user ID was rated as "bad"
5. number of banned IDs traded with user ID
6. data table of traded partners
7. log data

First, the rating value of the user ID score is given by trading partner in Yahoo! Auction. This score is assigned 1-point values ranging from -2 to +2. The first item is the total value. Second, the rating value average of trading partner is the total rating value to user IDs divided by trading partners. In other words, the value expresses

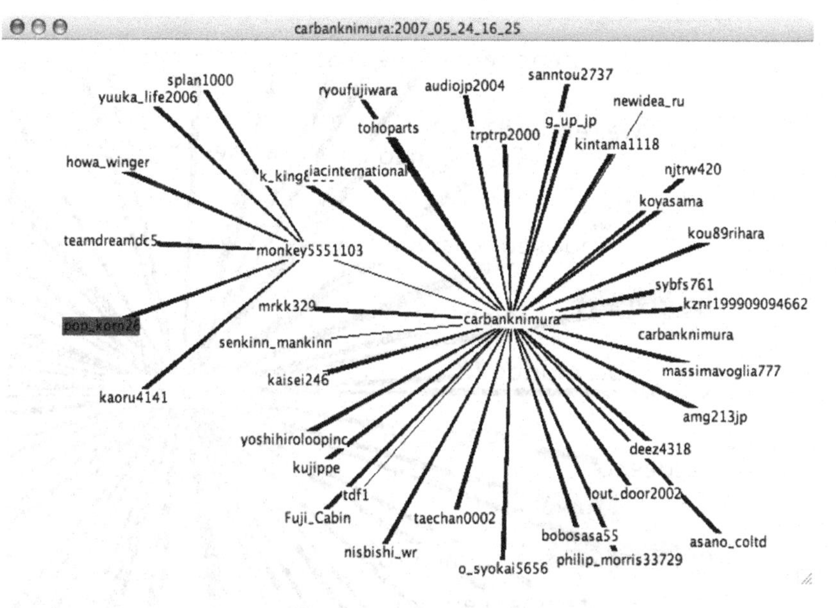

Fig. 3. Multi-node graph

the height of the grade of a trading partner of the user ID. Next, auction type represents that part occupied by seller or buyer user IDs in all transaction records as a percentage. For example, if there have been ten dealings, and six times the bidder was successful, the auction type is successful, and its numerical value is 60%. The percentage of the user ID was rated as "bad" is the number of user IDs have been rated "bad" or "very bad" divided by all partners. This value may be calculated highly even if rating value of user ID, the first item, is high. If a value of the fifth item is high, the user ID is suspicious. Next, the number of banned trading partners represents IDs banned by Yahoo or those lapsed for any reason. Some cheats have multiple disposable IDs. In the generated graph, banned ID nodes are indicated in red, as in Figure 4. In a precise sense, a banned ID cannot identify bad users such as frauds. The kinds of banned IDs can be estimated [7]. Next, the data table of trading partners includes much partner data such as ID names, rating value, number of transaction times, rating to user IDs, and status like seller or buyer. Finally, the log data is a list of files that were output when visualizing the graph in this system.

The above items are displayed in one window that also has a function that switches the view of the graph. There are three types of view: all view, seller angle, and buyer angle. All view displays every transaction relationship, including both seller and buyer stances of user IDs. Seller angle displays the seller stances of user IDs with all wide edges. Buyer angle displays the buyer stances of user IDs with all thin edges. Figure 5 is an output window that queries with certain user IDs. In addition, whenever queries are made with certain IDs, the Java program takes logs with the present time. In the log data window, log files are listed in a text area. Figure 6 presents log data with user ID "tdf1" in the window. The log file's name consists

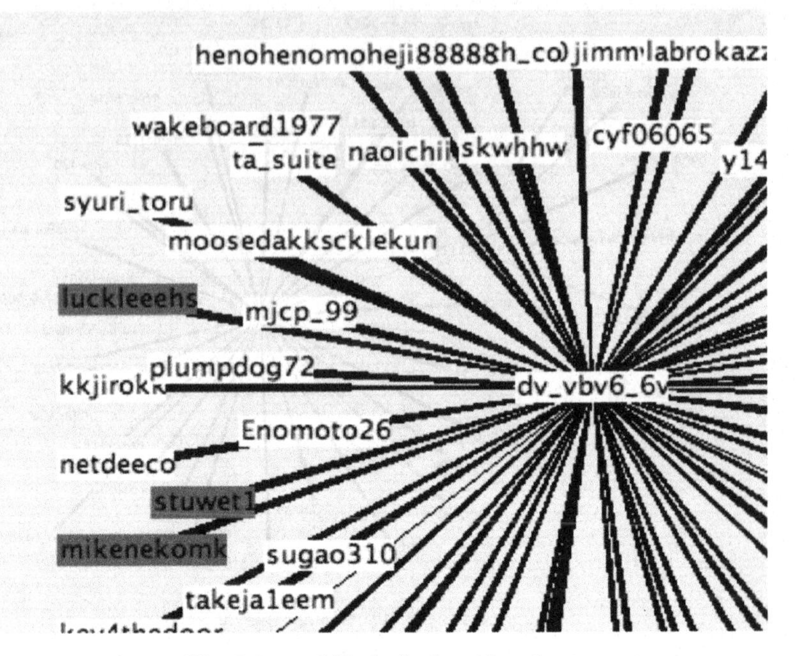

Fig. 4. Banned IDs is displayed in red

of the time it was generated and a user ID. Log files contain the above six items, ID name, and written time.

The data window assembles and displays the selected data. We created a Callback class by practically extending the HTMLEditorKit.ParserCallback to analyze HTML sources and sort this data in Java program. This class chooses HTML tags from the source code and only stores data that we can use for analysis in a table made with JTable class. The log data is one of the panels in the JTabbedPane on this window, and another is the data tab.

4 Preliminary Experiment

We can use the reliability of user IDs generated by this system as evaluation elements. However, such evaluation remains to be seen because the present phase is experimental. Therefore, we verified response time as an evaluation method. The following are the computer specifications used in our experiment. The operating system was a Mac OS X version 10.4.9. The processor was 2.16 GHz with Intel Core 2 Duo with 1 GB of memory.

This system has functions that display the time required from querying user ID as graph seed that is a word to become the basis of a search to graph beginning drawn on the window. The time depends on ID search in Yahoo! Auction [29]. We

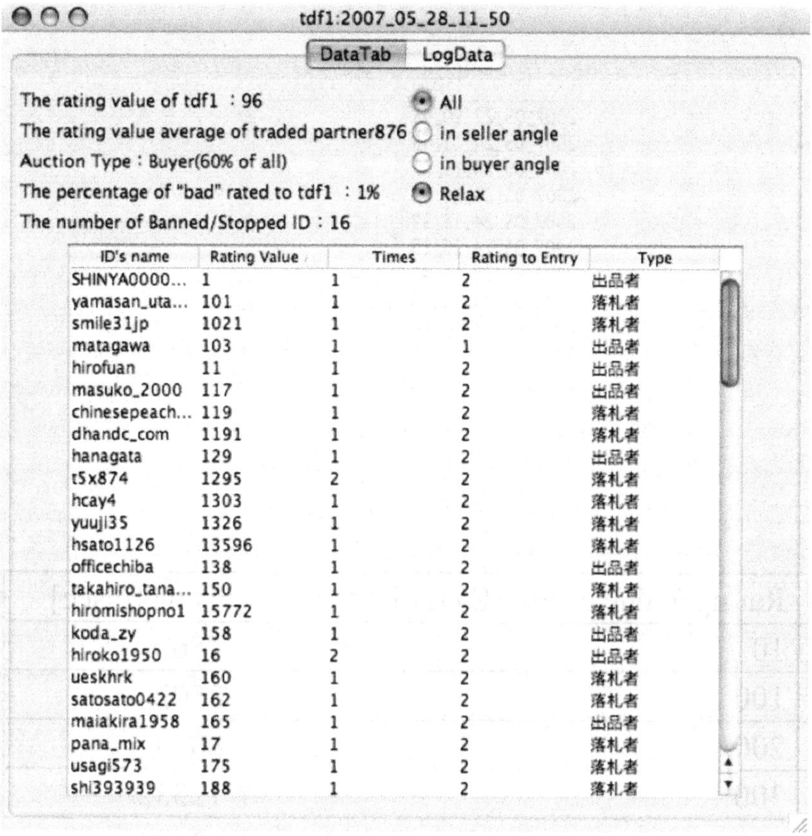

tdf1:2007_05_28_11_50

DataTab LogData

The rating value of tdf1 : 96
The rating value average of traded partner876
Auction Type : Buyer(60% of all)
The percentage of "bad" rated to tdf1 : 1%
The number of Banned/Stopped ID : 16

All
in seller angle
in buyer angle
Relax

ID's name	Rating Value	Times	Rating to Entry	Type
SHINYA0000...	1	1	2	出品者
yamasan_uta...	101	1	2	落札者
smile31jp	1021	1	2	落札者
matagawa	103	1	1	出品者
hirofuan	11	1	2	出品者
masuko_2000	117	1	2	出品者
chinesepeach...	119	1	2	落札者
dhandc_com	1191	1	2	落札者
hanagata	129	1	2	出品者
t5x874	1295	2	2	落札者
hcay4	1303	1	2	落札者
yuuji35	1326	1	2	落札者
hsato1126	13596	1	2	落札者
officechiba	138	1	2	出品者
takahiro_tana...	150	1	2	落札者
hiromishopno1	15772	1	2	落札者
koda_zy	158	1	2	出品者
hiroko1950	16	2	2	出品者
ueskhrk	160	1	2	落札者
satosato0422	162	1	2	落札者
maiakira1958	165	1	2	出品者
pana_mix	17	1	2	落札者
usagi573	175	1	2	落札者
shi393939	188	1	2	落札者

Fig. 5. Data window

experimented with several rating values of user IDs. This point expresses an identical item as the first value of the above data table. Table 7 represents experimental results.

The response time in this system grows almost linearly with the rating value of user IDs because the rating value of user IDs indicates the actual number of transaction partners. In Internet auction sites ratings after trading is customary, but not mandatory. The results are shown in Figure 8. Errors in the table reflect that it takes time to search the transaction records of several trading partners in the data table. In addition, the above response time is before a graph begins without regarding convergence time by swinging the Spring-Model.

AQ: I
the ci
Figur

5 Discussion

5.1 Advantages

The system we present has the following three advantages:

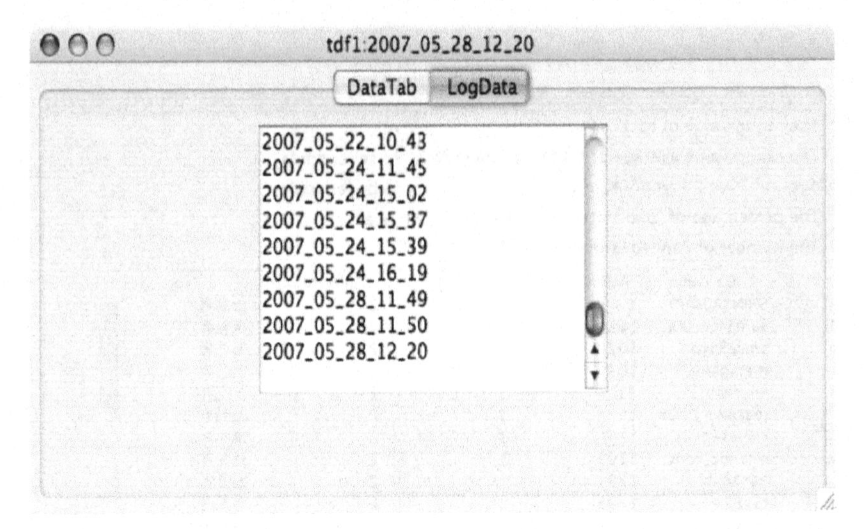

Fig. 6. LogData on window

Rating Value of user ID [point]	Time [msec]
10	456
100	3502
200	8050
300	12332
400	12624
500	15854
600	18069
700	22079
800	27723
900	32186
1000	34000

Fig. 7. Design features of the website versus the source of the site

1. Supports identification of cheats from transaction relationships.
2. Visualizes transaction relationships as network structure in various viewpoints.
3. Markets expectations with visualized network data.

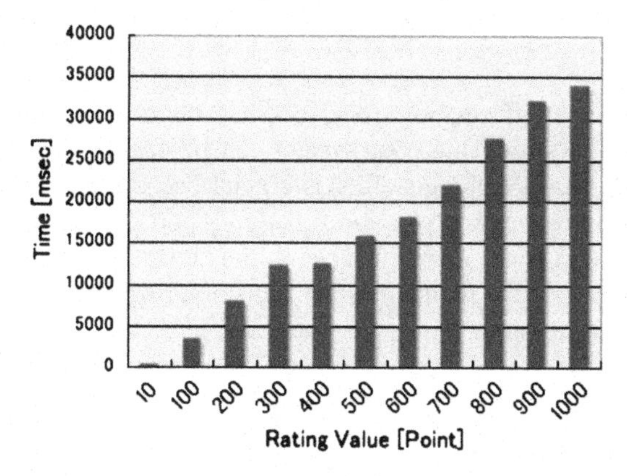

Fig. 8. Linear response time in system

The banned ID display function in the present system relates fraud detection in Internet auction networks. Such personal data as rating value and banned transaction partner ID is available for fraud detection. In the research area of large-scale databases, the fraud community is detected by mining rating time [7]. Here, they focus on fraudulent ratings for fraudulent acts that have dealings and get user trust with specified IDs. They represent that the community between frauders and fraudulent raters with this bipartite graph. They can be extracted from the dataset in Internet auction sites. This idea is included in our system. The log data of transactions is used to detect clues of fraudulent acts. On other hand, we can read characteristics from the visualizing graph by analyzing the auction network structure. Google [18] uses the PageRank algorithm to arrange search results. PageRank calculates value with a proving matrix eigenvalue problem that can also be applied to transaction relationships. We can handle hyperlink relations in a website as transaction relationships in Internet auctions. Also, web pages are treated as auction users.

In our second point, we aim to visualize the entire Internet auction network from transaction records with our system, but now we can only search transaction records with user IDs with auction search. A visualization model in this system would become increasingly complex with additional data. Therefore, selecting only useful data can reduce complexity. Other advantages of just exploiting use-related high or characteristic data are that system users can monitor auction networks from various visualization views.

Presently, market prediction has not yet implemented this system. The analysis of transaction relationships and relations between persons and items enables market prediction. Generally it is said that recommendation and a web personalization are not effective, but in late years the long tail attracting attention points out the effectiveness. The long tail is a law that a total of a low element of occurrence frequency holds a ratio that cannot be ignored for the whole. This law is applied in Internet auction sites.

5.2 Generality

Our system can extract information from web sites, but not from the inner databases managed by auction sites. Thus, our system can be easily modified to analyze a variety of auction sites. Such generality is crucial because currently many market systems are available, for example, ad-exchanging portals, insurance markets, etc. On such web-based markets, there must be trading networks. Fortunately, such web-based markets tend to have reputation mechanisms in which participants are rated based on particular measures. Thus, our methodologies can be applied to such web-based markets.

6 Summary

In this paper, we built a system that can extract transaction relationships on Internet auction sites that are visualized as graphs. The proposed visualization system has the following three advantages:

1. Supports the identification of cheats from transaction relationships and detects suspiciousness with user ID information such as selling or buying histories.
2. Presents a transaction relationship as a network structure from various viewpoints. Users can view the network from various angles such as seller or buyer.
3. Possesses explanation functions about visualized networks and predicts auction markets.

These components have been implemented in Java programming language and for some web-based scripts. As future works, we will inspect a method of visualizing auction networks with both personal user information and network structure. Most auction sites have a reputation mechanism in which each participant is rated based on its own activity in past trading. Although one study only used network structure for e-Bay trading [19], we will utilize reputation data to analyze trading network more effectively. Further we will develop a reasonable evaluation method and try a comparatively large-scale experiment.

References

1. Z. Gyongyi, H. G. Molina, and J. Pedersen, "Combating web spam with trustrank," In VLDB, 2004.
2. S. Brin and L. Page, "The Anatomy of a Large-Scale Hypertextual Web Search Engine," WWW7/Computer Networks, 1998, 30(1–7) pp. 107-117.
3. J. Kleinberg. "Authoritative source in a hyperlinked environment," In Proc. 9th ACM-SIAM Symposium on Discrete Algorithms, 1998.

4. K. Sugiyama and K. Misue: "Graph drawing by the magnetic spring model," J. of Visual Languages and Computing, vol. 6, no. 3, pp. 217–231, 1995.
5. http://auctions.yahoo.co.jp/jp.
6. http://auction.rakuten.co.jp/.
7. Y. Hirate, A. Akiyoshizawa, S. O, Y. Ioku, F. Kido, and H. Yamana, "Support System for Detecting Abuse Users in Internet Auction," DBSJ Letters vol. 5, no. 2, 2006.
8. H. Kashima, "Survey of Network Structure Prediction Methods," TJSAI vol. 22, no 3, pp. 344–351, 2007.
9. http://www.yukan-fuji.com/archives/2005/06/post_2487.html.
10. http://www.courts.go.jp/yamaguchi/about/osirase/20060201.html.
11. http://ascii24.com/news/i/topi/article/2000/04/14/608380-000.html.
12. P. Resnick, R. Zeckhauser, E. Friedman, and K. Kuwabara. Reputation systems. Communications of the ACM, 43, 2000.
13. P. Resnick, R. Zeckhauser, J. Swanson, and K. Lockwood. The value of reputation on ebay: A controlled experiment, 2003.
14. M. Melnik and J. Alm. Does a seller's ecommerce reputation matter? Evidence from ebay auctions. Journal of Industrial Economics, 50: 337–49, 2002.
15. http://sagi.sakura.ne.jp/rejume1.html.
16. Getoor, L. and Diehl, C. P.: Link mining: a survey, SIGKDD Explorations, Vol. 7, No. 2, pp. 3–12 (2005) .
17. P. Resnick, N. Iacovou, M. Suchak, P. Bergstrom, and J. Riedl. "GroupLens: Open Architecture for Collaborative Filtering of Netnews," In Conference on Computer Supported Cooperative Work, pp. 175–186 (1994).
18. http://www.google.co.jp/.
19. Shashank Pandit, Duen Horng Chau, Samuel Wang, Christos Faloutsos, NetProbe: A Fast and Scalable System for Fraud Detection in Online Auction Networks, World Wide Web, 2007.
20. Clarke, E. H. 1971. Multipart pricing of public goods. Public Choice 11:1733.
21. Groves, T. 1973. Incentives in teams. Econometrica 41:617–631.
22. Matsuo, T.; Ito, T.; Day, R. W.; Shintani, T. 2006. A Robust Combinatorial Auction Mechanism against Shill Bidders, In Proc. of the Fifth International Joint Conference on Autonomous Agents & Multi Agent Systems (AAMAS-2006).
23. Matsuo, T.; Ito, T.; Day, R. W.; Shintani, T. 2006. A Two-Stage Robust Combinatorial Auction Mechanism against Falsename Bids In Proc. of the IEEE Pacific Rim International Workshop on Electronic Commerce (IEEE-PRIWEC 2006).
24. Vickrey, W. 1961. Counterspeculation, auctions, and competitive sealed tenders. Journal of Finance XVI: 8–37.
25. Yokoo, M.; Sakurai, Y.; and Matsubara, S. 2004. The effect of false-name bids in combinatorial auctions: New fraud in Internet auctions. Games and Economic Behavior 46(1): 174–88.
26. http://www.npa.go.jp/cyber/statics/h18/pdf35.pdf.
27. http://en.wikipedia.org/wiki/Spoofing.
28. http://en.wikipedia.org/wiki/Phishing.
29. http://auctions.yahoo.co.jp/jp/show/searchoptions?catid=0.

A Cooperation Dealing Model of Hybrid Traders based on Volume Discounts

Satoshi Takahashi and Tokuro Mastuo

Yamagata University,
4-3-16, Jonan, Yonezawa, Yamagata, 992-8510, Japan.
takahashi2007@e-activity.org, matsuo@yz.yamagata-u.ac.jp

Summary. This paper proposes a new cooperation business model in which hybrid traders exist. We define hybrid traders as new traders on the Internet. Hybrid traders can become both buyers and sellers. We assume that hybrid traders do not have enough money. To buy items cheaply, hybrid traders cooperate with other traders. In regard to buying items, we consider a volume discount-based trading. We propose a mechanism in which trader cooperates, buys in a lot of goods, and increases own utility. Our mechanism adopts side payment to promote increasingly cooperation with traders. Cooperative traders commit participation based on a value of side payment. We extend mechanism which hybrid traders deal with multiple items. This mechanism shows new decision of side payment and proposer's strategy.

1 Introduction

In resent years, as e-commerce is developing, researchers regard e-commerce as very important subject of researches, such as, auction [1] and group-buying [4][5][6]. End-users can become both buyers and sellers since it is easy for them to open their shops on the web. Such transaction is called B2B(Business to Business)/ B2C(Business to Consumer) [8]. Generally, when end-users open shops on the web, it takes less cost and money. It is easy for consumers to be sellers like a company. In this paper, we call hybrid traders such end-users. When hybrid traders sell items to general consumers, they need purchase items. Hybrid traders do not have enough money to get items. To purchase items at a low price, some traders cooperate with each other on the web since users can communicate with each other easily. Namely, they purchase items as joint capital. In this case, items are sold based on volume discount [10] from sellers such as producers, factories because cooperated traders can purchase a lot of items in one time. If a trader has enough budgets and he/she can purchase a lot of items, price of each item goes down. Although each trader does not enough money, they can purchase in items at a lower price making purchasing community.

When traders cooperate with each other, it is possible to purchase cooperatively cheaper than individually. As the result, each trader's utility increases. If all traders know about types of traders, they make cooperation easily.

S. Takahashi and T. Mastuo: *A Cooperation Dealing Model of Hybrid Traders based on Volume Discounts*, Studies in Computational Intelligence (SCI) **110**, 101–111 (2008)
www.springerlink.com

We employ side payment mechanism to promote trader's cooperation. If traders are rational, all traders must cooperate with each other. Each trader has a certain participation incentive based on a valuation of side payment. Side payments are given as cooperation fee from proposing traders to cooperations. If the former's utilities decrease paying side payments to cooperative traders, none search for cooperative agents. This paper proposes a mechanism in which traders' utilities are becoming maximum searching for optimal side payment value. We also propose a decision method of utility maximization and, a mechanism of optimal budget allocation in dealing multiple items.

The rest of this paper consists of the following eight parts. In section 2, we define several terms, assumptions and method of items allocation. In section 3, we define hybrid traders. Then in section 4, we define side payment. In section 5, we propose a mechanism of single item dealing. In section 6, proposing mechanism for expanding multiple items dealing. After that, we discuss about expanding mechanism. Finally, we present our conclusion.

2 Definitions and Assumptions

In this section, we give some definitions of terms and assumptions in our proposed mechanism.

$H = \{h_1, ..., h_i, ..., h_n\}$: A set of hybrid trader with participation of web community.
$A = \{a_1, ..., a_j, ..., a_m\}$: A set of tradable items. All items are sold based on volume discount.
$B = \{b_1, ..., b_i, ..., b_n\}$: Budgets of hybrid traders.
$v_j(\omega)$: A price of item a_j when purchasing with volume discount of the ω phases. ω is parameter that shows number of items which implement discounts.
$p_{i,j}$: A price in which hybrid trader h_i sells item $a_j$for end-users.
$U_i = p_{i,j} - v_j(\omega)$: A utility when hybrid trader h_i sells item a_j. $(0 \leq U)$
$S_{i,n}$: A value of side payment that hybrid trader h_n pays h_i. $0 \leq S_{i,n} \leq U_n - U'_n$ (U'_n: Utility of h_n with independent transaction)
$Q_{i,j}$:Number of items in which hybrid trader h_i buys in item a_j.

Definition 1 Hybrid traders participate web site community. Hybrid traders deal in this community.
Definition 2 Hybrid trader can purchase all items restricted budgets.
Definition 3 Hybrid traders propose cooperation of buying-in for other hybrid traders.
Assumption 1 Hybrid traders do not have enough money. They do not have enough budgets in which hybrid traders get a grace of volume discount.
Assumption 2 All items are sold as volume discount. Hybrid traders know price and discount ratio of items.
Assumption 3 Items are sold with hopeful price. There are no risk of dealing.

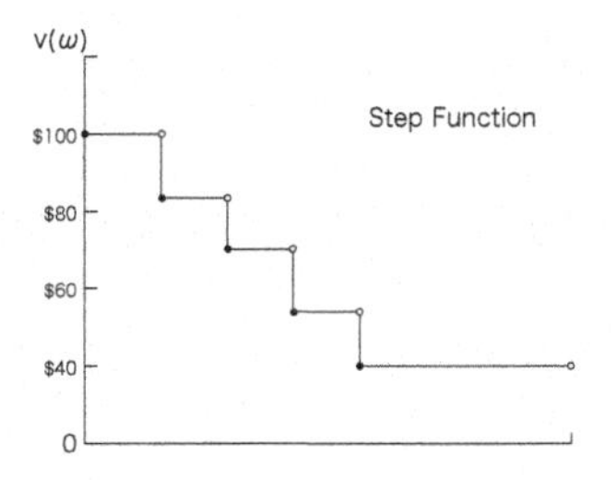

Fig. 1. Volume Discount

Table 1. Value of Item

Quantity	$
1 - 10	100
11 - 50	85
51 -100	70
101 - 200	55
201 -	40

2.1 Volume Discount and Allocation of Items

On above assumption, all items are sold with volume discount. The item price is cheaply by number of items. Figure 1 shows stair-case graph indicating items price in volume discount. It shows prices of items are step function. Increasing number of items, price of each item goes down. Table 1 shows a concrete example of Figure 1. When traders can purchase 11 items, utility calculates on more increasingly about $15 when they purchase only one item.

Hybrid traders sell items with their own gains to end-users. In assumption 1, hybrid traders can not treat a lot of items because they do not have enough money. However, if hybrid traders can cooperate and trade efficiently, they get opportunity of increasing utilities by volume discount. Then, how the items are allocated ?

Hybrid traders allocate the items based on percentage of investing. Total number of allocated items $\sum_{i=1}^{n} Q_i$ equals total number of items bought-in. Figure 2 shows items allocation. Hybrid trader h_i allocates items in total number of items by ratio of investment b_i on total investment $B = \sum_{i=1}^{n} b_i$.

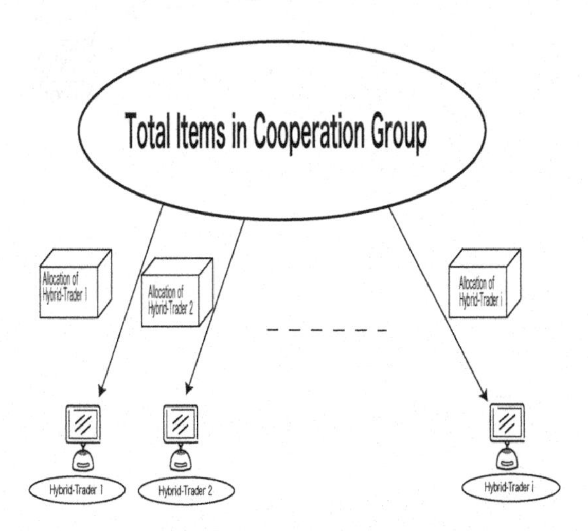

Fig. 2. Allocation of Items

3 Hybrid Trader

In this section, we define hybrid traders. On economic phenomena, we treat dealing between sellers and buyers. But in under continual time, same people sometimes play the seller and the buyer. And the people are end-users who only buy items basically. The sellers are special people who have a certain level of money and procedure. It is difficult to be sellers in which they pay stored cost and advertisement cost without enough money to spend in trading. However, economic activity on the Internet is no cost of their payment. Additionally, users learn indirectly about selling procedure because they use auction and group buying.

End-users do not have enough money for buying-in a lot of items. Traders who have little money can purchase by a pool of capital. One of characteristics of the Internet is that traders can cooperate many and unspecified people. In case of items are sold with volume discount, traders can purchase more items by same budget because a unit price of each item becomes a discounted price.

We define a user who plays seller and buyer as hybrid trader. Figure 3 shows model of transaction environment when hybrid trader stands on their environment. The seller only sells items. The buyer also buys items. Hybrid trader can sell and buy items.

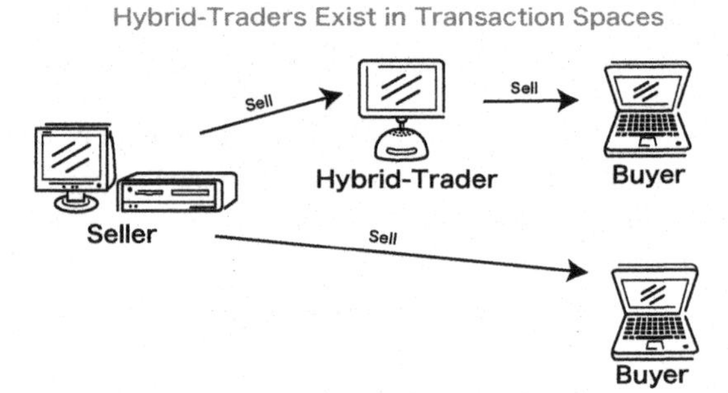

Fig. 3. Hybrid Traders

4 Side Payment

In this section, we define side payment institution. We adopt side payment institution for incentive of cooperation to proposer. Side payment is monetary transfer based on some kick-back for cooperation hybrid traders. Incentive of cooperation increases by it. Vendors who pose bid collusion pay kick-back for incentive of cooperation and dropping out of bit collusion. Cooperators can get items cheaply by cooperation of purchasing items. U_i is defined as hybrid trader h_i's utility. U is difference between payments and valuations, such as, $p - v_i(\omega)$. As show in assumption 1, hybrid traders do not have enough money to get items of themselves. In case of existing cooperation traders, a proposer purchases items more cheaply by increasing budget. The proposer's utility is calculated as $U' = p - v_i(\omega)$. His/her utility increases as $U - U' \geq 0$. Consequently, side payment should be paid between 0 to $U - U'$.

5 Single Item Dealing

We handle a situation of single item dealing with value of side payment.

5.1 Cooperation Negotiation which Depends on Side Payment

All hybrid traders have participation probability depending value of side payment. Hybrid traders are separated into some sets $T_l : \{l = 1, ..., 2, ..., l\}$ by own participation probability. The probability of set T_l is described by function $f_l(s_i)$. s_i is ratio of side payment in which hybrid trader h_i decides. Cooperators decide to participate in purchasing group by that the proposer shows side payment. When value of side payments are increased, proposer h_i's utility reduces due to paying side payment. Instead of this, many cooperators join in the purchasing group. In this case, proposer's utility decreases. If we decide optimal value of side payment, proposer's utility is maximized.

5.2 Dealing Procedure

We shows procedure of single item dealing.

step 1 Hybrid trader h_i is a proposer. h_i proposes about cooperation of purchasing items a_j for other hybrid traders. All traders know discount ratio of items.
step 2 The proposer h_i shows ratio of side payment as s_i.
step 3 Other traders commit participation by side payment.
step 4 The proposer h_i gathers money from purchasing group and purchases the items.
step 5 The items are allocated by each contribution.
step 6 Each trader sells the items by own accountability.
step 7 The proposer pays side payment to all cooperative traders. In this payment, the proposer pays $s_i \cdot (U - U')$ with contribution. A payment $S_n = \{s_i \cdot (U - U')\} \cdot b_n / B$

5.3 Optimization Side Payment

Hybrid traders have participation probability[11] depending on value of side payment. Here, we set up an assumption.

Assumption 4 Proposer knows participation probability.

Quantity	$
1 - 5	7
6 - 10	5.5
11 -	3

Table 2. Value of Item 1

A proposer can optimize a value of side payment adopting this assumption.

We consider that there are two hybrid traders and a tradable item. Trader h_1 proposes to the other trader cooperating on item a_1. Item a_1 has 3 levels of discount rates. Table 2 shows the item price based on number of items. Trader h_1's budget b_1 is $30. If no traders cooperate, trader h_1 buys 4 items for $28. If trader h_2 who has $3 cooperates with h_1, total budget because $33. Traders h_1 and h_2 buy 11 items and pay for $33. We assume that item's price, where the item is sold to end-users, is $10. Range of side payment is $0 \leq s_1 \leq 4$ per one item.

Here, we show a formalization of optimization. Hybrid trader h_1 is the proposing trader. There are n traders in purchasing community and m items. In this condition, trader h_i proposes purchasing item a_j. Items are sold by volume discount. Item's price is $v_j(\omega)\{\omega = 1, ..., \omega', ..., \omega\}$. $v_j(\omega)$ is the cheapest value. Number of l group $T_{\{1,...,2,...,l\}}$ exists with depending probability. All participants reside it. Group $T_{\{1,...,2,...,l\}}$ has participation probability $f_{\{1,2,...,l\}}(s_i)$ depending on side payment. If the proposer gathers n cooperators, he/she decides cooperators based on $n = \sum_{l=1}^{l} f(s_i) \cdot T_l$.

It is possible to calculate the number of traders shown as this formula. The proposing trader uses this formula and decides an optimal value of side payment. When proposer pays side payment, value is $s_i \cdot (U - U')$. U is utility $U_i = \{p_{i,j} - v_j(\omega)\} \cdot Q_{i,j}$ when he/she cooperates. U' is utility of individual dealing.

5.4 Simulation

We simulate based on preceding definition and assumption. We consider hybrid traders who have 3 types of preferences like Figure 4. Figure 4 shows participation probability about changing side payment among $0\% \leq s \leq 100\%$. Type 1 is a group of hybrid traders who has preference which is participation probability rising nonlinearity. Type 2 is a group who has incentive to participate near 0.5. Type 3 is hybrid traders who have participation probability rising linearity. If the value of side payment grows, the cooperator gets less side payment. When the value of side payment is just 0.5, traders who are classified in type 2 participate in cooperation. Table 2 shows concrete values of hybrid traders' utilities. Each trader has budget between $20,000 and $200,000. We change ratio of side payment like $0\% \leq s \leq 100\%$. We set up his/her budget based on uniform distribution. Figure 5 shows a visual comparison between proposer's and cooperator's utilities. Table 3, cooperators' utility comes back proposer's utility when investment between $50\% \leq s \leq 60\%$. In this result, the best value of side payment is among $50\% \leq s \leq 60\%$.

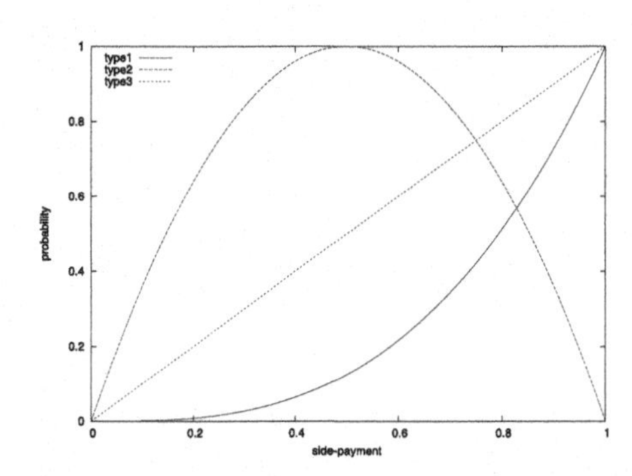

Fig. 4. Type of Hybrid-Traders

Table 3. Utility

side-payment	proposer($)	cooperator($)
0.0	139	0
0.1	12326	9015
0.2	12631	9281
0.3	12249	9364
0.4	11435	9424
0.5	10274	9478
0.6	8800	9413
0.7	7051	9500
0.8	5054	9482
0.9	2513	9552
1.0	155	9486

Assumption 4 is important condition in this simulation. But, it is difficult to know other traders' type. We propose a method that value of side payment is decided mechanically without assumption 4. Side payment is decided not by depending on side payment but by based on merit of cooperation. Proposer might not gather enough money. But if negotiation communities are made by types of purchasing items, more traders cooperate to purchase items. We separate hybrid traders by their types and create small groups. We define the following definition.

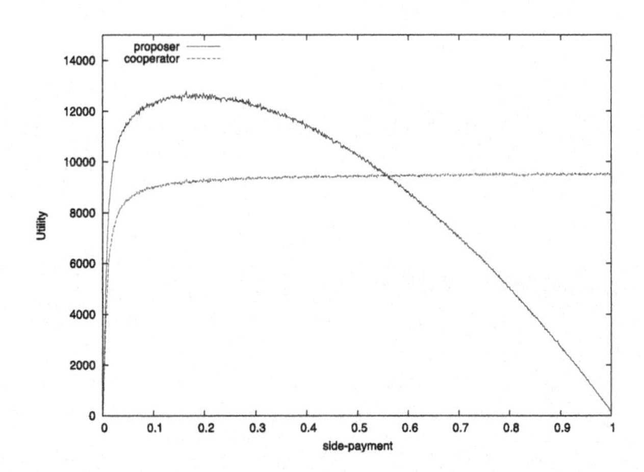

Fig. 5. Result of The Simulation

Definition 4 Hybrid traders who use e- commerce site employing our mechanism must propose purchasing items.

Hybrid traders can purchase multiple items on this definition.

6 Multiple Items Dealing

We aspire for fair allocation of payoff by deciding appropriate side payment mechanically. In this section, we consider about a case where a hybrid trader purchases multiple items. Purchasing group should be small set, since dealing in large-scale group is complication and makes computers take a lot of costs to compute the allocation.

6.1 Additional Definition and Protocol

We show an additional definition on dealing multiple items.

$G = \{G_1, ..., G_k, ..., G_l\}$: G is set of hybrid traders. Hybrid traders join a small group. When hybrid trader h_i joins in G_k, h_i do not join in other groups. The following equation shows total sets of groups when there are l groups. $H = \sum_{k=1}^{l} G_k$.

$G' = \{G'_{1,1}, ..., G'_{k,j}, ..., G'_{l,m}\}$: $G'_{k,j}$ is a set in which traders purchases item I_j in small group G_k. ($G'_{k,j} \subseteq G_k$).

Here we show a protocol in trading among multiple hybrid traders on many items.

Protocol

- Hybrid traders are separated by type of purchasing. Hybrid traders have preference about dealing items.
- All hybrid traders in a purchasing group must propose purchasing item.
- If there are no cooperator, item are not dealt with traders.
- Side payment is paid after items are allocated.

Further, a proposer has two strategies of cooperation.

1. Cooperating with each other.
2. Using side payment institution.

- In strategy 1, side payments are not paid.

Hybrid traders in group must propose purchasing items. Everyone can cooperate on those items. When a trader does not want to pay side payment, they take cooperation instead of paying it.

We define formula of determine from of side payment mechanically as following formula.

$$S_{i,n} = \{\underset{i \notin G'_j}{v_j(\omega')} - \underset{i \in G'_j}{v_j(\omega)}\} \cdot Q_{i,j}$$

Our mechanism calculates differences between purchasing price in case that when hybrid trader h_i participates cooperation group and purchasing price in case that hybrid trader h_i does not participate cooperation. The mechanism also computes multiplication number of purchasing item by h_i ($\omega' \leq \omega$). This method can restrain that participants who joins in a cooperating group when purchase price is minimum.

We show proposer's strategy. We adopt not only side payment but also kick-back which increases utility. Figure 6 shows an example of proposer's strategy. $Strategy1$

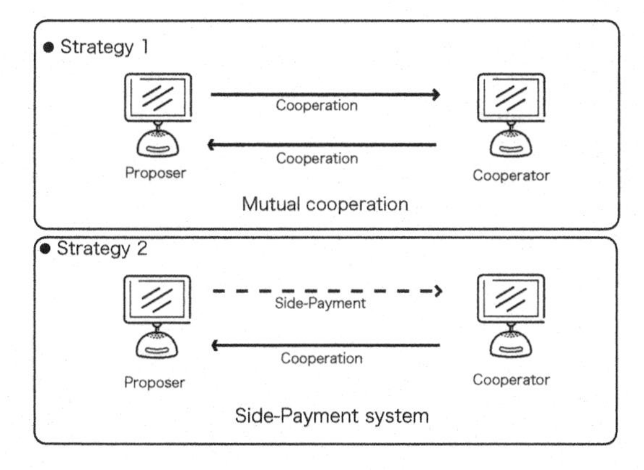

Fig. 6. Strategy of Proposer

Table 4. Value of Items

Quantity	a_1	a_2	a_3
1 - 10	\$100	\$80	\$50
11 - 20	\$70	\$60	\$40
21 - 50	\$50	\$45	\$30
51 -	\$35	\$30	\$20

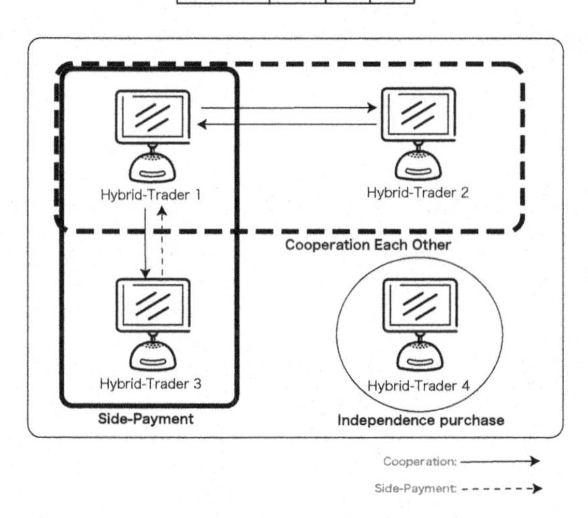

Fig. 7. Cooperation Negotiation with Small Group

is a strategy of cooperation with each other. Proposer takes not to pay side payment but to cooperates against cooperator. Side payment is not occur. $Strategy2$ is a side payment institution. Proposer purchases items more cheaply and aspires increasing utility by using two strategies.

It is difficult for free riders to increase utility by deciding side payment automatically. Proposing by all hybrid traders can restrain free rider.

7 Discussion

7.1 Example of Multiple Items dealing

We shows an example of multiple items dealing. Four hybrid traders H=h_1, h_2, h_3,h_4 including in group G_1. Three items $A = \{a_1, a_2, a_3\}$ are tradable. Table 3 shows each price and volume discount prices. Each first price of items is $v_1(1) = \$100$, $v_2(1) = \$80$, $v_3(1) = \$50$. Each hybrid traders budget is $b_1 = \$1,000$, $b_2 = \$700$, $b_3 = \$500$, $b_4 = \$600$. Proposing items of each trader are $h_1 : a_1, h_2 : a_1, h_3 : a_3, h_4 : a_2$. Figure 7 shows cooperation and collateral after negotiation among them. Hybrid traders h_1, h_2 propose same items. They cooperate with each other. h_1 cooperates with h_3 and pays $180. h_3 selects side payment against cooperating with h_1. h_4 does not purchase when item's price is more expensive than his/her budget because no one cooperates. Hybrid trader h_1 gets 16 of item a_1 and 3 of item a_3. Hybrid trader h_2 gets 14 of item a_1. Trader h_3 gets 8 of item a_3. Each purchasing prices is $p_1 = \$120$, $p_2 = \$100$, $p_3 = \$60$. h_1's utility is $1240. h_2's utility is $980. h_3's utility is $320. In this transaction, h_1 gets side payment from h_3. Side payment value is $S_{1,3} = (\$80 - \$60) \cdot 3 = \$60$. h_3 pays $60 to h_1. It is trivial to increase hybrid traders' utilities. To buy multiple items are more increasingly than to buy single item. It is important to allocate budget to buy multiple items.

In multiple items dealing, allocation of budget is one of important problems regarding as volume discount-based trading. Purchasing multiple items is more increasingly of utility than purchasing only one item. A simple method for allocation of budgets is that the mechanism divides number of items from total budgets and prepares each divided budget to purchase items. However, the above mechanism sometimes allocates inappropriate items where users' payoffs decrease. Thus, we consider generalized trading of combinatorial items and budgets where users' utilities are maximization/semi-maximization.

8 Conclusion

This paper shows a mechanism of dealing with hybrid traders who play both sellers and buyers. We defined cooperation dealing that hybrid traders cooperate with each other. We also defined side payment as incentive of cooperation. In single item dealing model, we proposed a method of deciding optimal side payment with restricted assumption. In multiple items dealing, we proposed a method of decision side payment mechanically. We showed restraining decreasing proposer's utility and fee rider problem in the cheapest price by decision of kick-back. Our future work includes decision method of the best side payment transfer method without assumptions, combinatorial items and budgets, and a method of restraining free rider completely.

References

1. K. Leyton-Brown, Y. Shoham, and M. Tennenholtz, Bidding clubs: institutionalized collusion in auctions, Proceedings of the 2nd ACM conference on Electronic commerce, pages 253–259, 2000.
2. T Matsuo, T. Ito and T Shintani, A Discovering Method of Shill bidders in Combinatorial Auctions, The 19th Annual Conference of the Japanese Society for Artificial Intelligence, 2005.
3. Tokuro Matsuo and Toramatsu Shintani, An approach to detecting bid-rigging in auction Japanese Society for Artificial Intelligence, 21(5):B16.1–B16.8, 2006.
4. J. Yamamoto and K. Sycara, A stable and efficient buyer coalition formation scheme for e-marketplaces, Proceedings of the fifth international conference on Autonomous agents, pages 576–583, 2001.
5. H Ito, T. Ochi and T. Shintani, A Group Buying Protocol based on Coalotion Formation for Agent-mediated E-Commerce, International Journal of Computer and Information Science(IJCIS), 2002.
6. C. Li and K. Sycara, Algorithm for combinatorial coalition formation and payoff division in an electronic marketplace, Proceedings of the first international joint conference on Autonomous agents and multiagent systems: part 1, pages 120–127, 2002.
7. Tokuro Matsuo, A New Pooled Buying Method Based on Risk Management, IEA/AIE 2007, LNAI 4570, pages 953–962, 2007.
8. E. Turban, J. Lee, D. King, and M. Chung, Electronic Commerce:A Managerial Perspective, Pearson Education, 2000.
9. T Matsuo, T. Ito and T Shintani, A volume discount-based allocation mechanism in group buying, International Workshop on Data Engineering Issues in E-Commerce (DEECf05), 2005.
10. Tokuro Matsuo and Takayuki Ito, A decision support system for group buying based on buyers' preferences in electronic commerce, In the proceedings of the Eleventh World Wide Web International Conference (WWW-2002), pages 84–89, 2002.
11. R. H. Varian, Intermediate Microeconomics A Modern Approach, volume seventh Edition, W. W. Norton and Company, 2005.

Analyses of Task Allocation based on Credit Constraints

Yoshihito Saito and Tokuro Matsuo

Department of Informatics,
Graduate School of Engineering,
Yamagata University,
4-3-16, Jonan, Yonezawa, Yamagata, 992-8510, JAPAN.
saito2007@e-activity.org
matsuo@yz.yamagata-u.ac.jp
WWW homepage:
http://veritas.yz.yamagata-u.ac.jp/

Summary. This paper presents a new contract model of trading with outsourcer agents and developer agents in large-scale software system manufacture. We consider a situation where ordering party does not order making software directly. Large-scale software consists of some modules. If the scale of a module is biggish, the software can be efficiently developed ordering as divided modules to some software developers. Generally, software developer has some risks as a company, such as, bankruptcy and dishonor. In such situation, it is important for an outsourcer to know how to reduce a rate of risks. In this paper, we propose a new risk diversification method of contracts with software developers in dividable software systems. In our protocol, we employ a payment policy of initial payment of and incentive fee. Then, the payment amount of initial fee is based on the developer's credit. Thus, our protocol prevents an outsourcer from risk on the project. Further we propose a distributed task model to reduce time of development. The results of experiments show the effective strategy for ordering party where risk and number of developers change. Our simulation shows that the outsourcer can get much earnings and performance selling/using the software at an early date when the number of modules and developers increase.

1 Introduction

In recent years, tradings using the Internet/computers develop [6]. Computer-based commerce is one of effective form of economic activities [2][4][7]. Instead of items trading, software can be ordered as tasks from outsourcers to developers electronically. Each trader is implemented as a software/autonomous agent. An outsourcer agent negotiates with developers on costs and quantities of tasks and decides price and allocations based on results of negotiations. He/she orders tasks to a developer agent. In this paper, we consider the situation that outsourcer orders tasks to many companies in software manufacture. Many software-developing companies increase

Y. Saito and T. Matsuo: *Analyses of Task Allocation based on Credit Constraints*, Studies in Computational Intelligence (SCI) **110**, 113–125 (2008)
www.springerlink.com

and they extend their business to receive a contract from corporate parent and some other companies. In general, most of large-scale software system consists of multiple modules and sets of classes. If such software is ordered to only one software developing company, it takes a lot of time and costs. If the software system can be divided as some middle sizes of modules, the outsourcer can order the system to multiple developers with divided modules. On the other hand, when there are some risks, such as bankruptcy and dishonor, outsourcers should consider how they can order to developers effectively. These are important to make decisions and strategy to win the competition against other software developers.

In this paper, we consider the effective strategy for the ordering party to save lots of time and costs. When outsourcer determine developers whom can perform task successfully and cheaply. The development company proposes concrete cost of his/her work and the outsourcer ask to discount of the price. Outsourcer company sometimes orders the task as a set to one developer. In this case, he/she considers only one developer's ability to perform the task. However, there is a certain risk since the developer goes out his/her business due to bankruptcy. It takes a lot of money and cost in this situation. Outsourcer needs much money to complete his/her project.

To solve the problem, we consider the divided ordering method avoiding the risk such as chain-reaction bankruptcy. First, developing companies evaluate a value for each module considering the scale of task. Then, they bid their valuations by sealed bid auction. The outsourcer calculates a minimized set of all development parties' valuations. In this protocol, it takes less cost even though developer stops his/her business. However, from standpoint of time, it may take a lot of time since a developer serves tasks more than two in same time. To solve the time problem, we consider the model where each task distributes to more developers. Time of development is reduced due to distributed task.

To compare and analyze with the above two situations, we give some simulation result for some conditions. Our experiment shows the relationship between the risk of bankruptcy and outsourcer's cost. In our simulations, when number of module increases, outsourcer should order as distribution. In less number of modules and developers, good strategy for outsourcer to reduce cost of ordering is that he/she orders to only one developer. There are less modules in software, outsourcer prevents high costs from the risk. On the other hands, when the number of modules and companies increase more and more, good strategy for outsourcer to reduce cost of ordering is that he/she orders to multiple developers distributionally. Further, in such situation, outsourcer reduces the development period to allocate the condition such that each agent serves only one task. By just that much, the outsourcer can get much earnings and performance selling/using the software at an early date.

The rest of this paper consists of the following six parts. In Section 2, we show preliminaries on several terms and concepts of auctions. In Section 3, we propose some protocols in distributional software manufacture. In Section 4, we conduct some experiment in situations where number of modules and companies increase. In Section 5, describes some discussions concerned with development of our proposed protocol. Finally, we present our concluding remarks and future work.

2 Preliminaries

2.1 Model

Here, we describe a model and definitions needed for our work. The participants of trading consist of an ordering party and multiple software developers. The outsourcer prepares the plan of order to outside manufacturers, and developers declare evaluation values for what they can serve the orders. The outsourcer orders the software development companies to do subcontracted implementing modules. We define that the cost is including developer's all sorts of fee and salary.

- At lease, there is one ordered software project. The architected software consists of a set of dividable module $M = \{m_1, \ldots, m_j, \ldots, m_k\}$. m_j is the jth module in the set.
- d_i is the ith contracted software developer with an outsourcer in a set of developers $D = \{d_1, \ldots, d_i, \ldots, d_n\}$.
- Software developers declare a valuation of work when they can contract in implementation of the modules. $v_{ij}(\ v_{ij} \geq 0)$ is the valuation when the developer d_i can contract for implementation of the module m_j.
- p_{ij}^{pre} is an initial payment for developer d_i paid by the outsourcer.
- p_{ij}^{post} is an incentive fee paid after the delivery of the completed module.
- v_{ij} is $p_{ij}^{pre} + p_{ij}^{post}$.
- Condition of software development company consists of his/her financial standing, management attitude, firm performance, and several other factors. The condition is shown as A_i integrated by them.
- The set of allocation is $G = \{(G_1, \ldots, G_n) : G_i \cap G_j = \phi, G_i \subseteq G\}$.
- G_i is an allocation of ordering to developer d_i.

Assumption 1 (Number of developers) *Simply, we assume $n > k$. There are lots of software developers.*

Assumption 2 (Payment) *There are two payment such as advanced-initial payment and contingent fee. Realistically, the former increases the incentives of making allocated modules. When the module is delivered successfully, the contingency fee is paid to the developer.*

Assumption 3 (Risks) *In the period of developing the modules, there is a certain risk r_i for developer d_i, such as bankruptcy and dishonor. We assume that r_i is calculate as $1 - A_i$.*

Assumption 4 (Dividable modules) *We assume that the large-scale software can be divided as some middle size modules.*

Assumption 5 (Integration of Modules) *We assume that some modules can be integrated without the cost.*

2.2 Initial Payment

When developers serve tasks from ordering company, a partial payment is paid before they start developing modules. In actual unit contract of software implementation, the payment sometimes divided as an initial payment and incentive fee. In this paper, we assume that v_{ij} is $p_{ij}^{pre} + p_{ij}^{post}$. In general, p_{ij}^{post} is sometimes increased based on the quality of finished work. However, simply, we do not consider it.

For example, the value A_i of condition of software development company is calculated based on his/her financial standing, management attitude, firm performance, and several other factors. If A_i is higher value, the developer has the credit. On the other hand, if A_i is near zero, the company does not enough credit. Some companies may have been just now established. If a company has enough credit, they need not do fraud since they have a steady flow of business coming in due to his/her credit.

In this paper, to make simple discussion, we assume that the developers must complete the performance on contract when they are winner of the auction. Concretely, we give the following assumption.

Assumption 6 (Performance on contract) *There are no developers cancel and refusal allocated tasks without performance on contract. Namely, the condition of participation to bidding is performance of business without performance of business.*

2.3 Contract

When an outsourcer orders an architecture of software to development vender company, there are mainly two types of trading. One is the trading by contract at discretion. Ordering company determines the developers ordering making software. They decide the price of the work. For example, the development company proposes concrete cost of his/her work and the outsourcer ask to discount of the price. Another type of trading is a policy of open recruitment. In this trading, the companies who can accept the order from outsourcer compete on price. In other words, developers who can serve the work selected by bidding, such as an auction. First, the outsourcer shows the highest value in which they can pay for the scale of software. When the cost for developers is less than the value, they declare to join in the bidding. In this paper, we consider the latter case of contracts.

When developers who participate in the competition, they give bid values for the task. For example, there are three developers d_1, d_2, and d_3. If the developer d_2 bids the lowest value in three developers, the developer d_2 contracts for the implementation of software ordered by the outsourcer.

Here, we show a simple protocol of contract. Figure 1 shows the following contract model. In this figure, developer 2 serves all tasks as a whole. Namely, developer 2 bids the lowest valuation comparring with other all developers. Developer 2 needs to complete all modules.

Protocol 1

1. For the large-scale software, an outsourcer offers for public subscription.

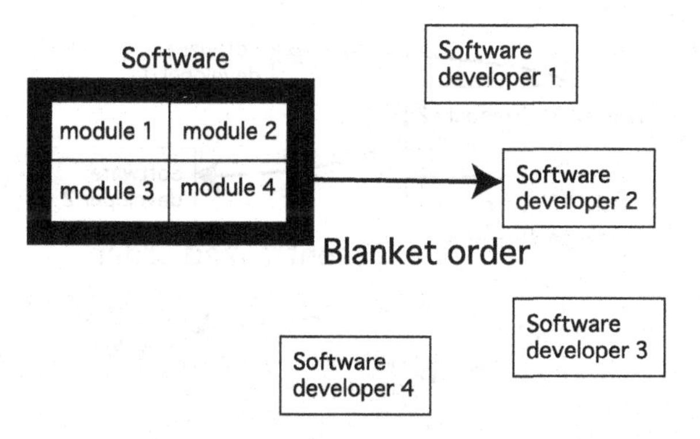

Fig. 1. Protocol 1

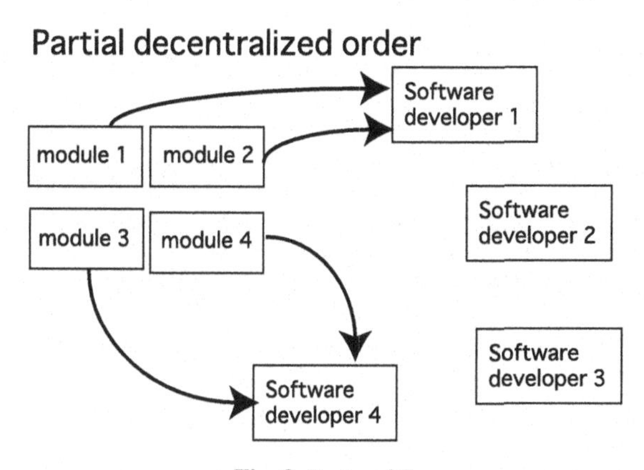

Fig. 2. Protocol 2

2. Software developers who can contract with the outsourcer come forward as contractor.
3. Developers submit a cost value of the task to the outsourcer by sealed bid auction. Namely, they bid $\sum_{j=1}^{k} m_j$.
4. A developer who bids the min $\sum_{j=1}^{k} m_j$ contracts with outsourcer for the declared cost.

For example, there are 3 developers. d_1 bids for 100, d_2 bids for 80, and d_3 bids for 50. In this case, developer d_3 serves the work for 50. Here, we consider that the developer 3 go out of business due to a certain factor. We assume that initial fees p_{ij}^{pre} of the work are paid as fifty percent of the contract prices. Namely, incentive fee p_{ij}^{post} is determined another fifty percent of cost. The outsourcer lost initial fee of developer d_3 for 25 dollars. The outsourcer orders and re-allocates the task to the developer d_2 since he/she bids the second lowest value. Totally, the ordering party takes 105 since it needs initial fee for developer d_3 and contract fee of developer d_2. In this contract,

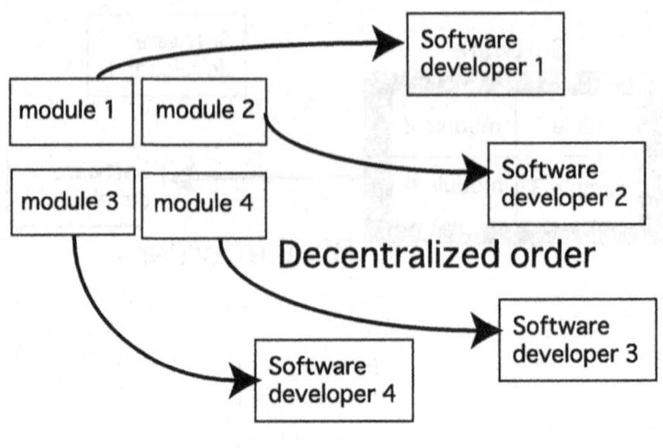

Fig. 3. Protocol 3

it is very high risk for the outsourcer if developer becomes bankruptcy during serving the tasks.

3 Protocol

In this section, we propose concrete protocol to determine the winner of contract. Here, we consider the risk about the developers. There are some risks for developers as a company, such as, bankruptcy and dishonor. To reduce the rate of risks, we propose a new diversification of risk based on divided tasks in large-scale software system manufacture. Further, we employ the advanced-initial payment and contingent fee as payment from an outsourcer. The former increases the incentives of making allocated modules. When the module is delivered successfully, the contingency fee is paid to the developer.

In actual trading, the above protocol 1 has a problem concerned with risks. After the outsourcer pays the initial payment, the developer who contracts with outsourcer starts implement software. However, the developer might get out of business and bankruptcy due to the problem of their company's financial problem, and other undesirable factors. When the developer declares his/her cost for five million dollars, the outsourcer pays the initial fee for two million dollars. If the developers become bankruptcy, the outsourcer lost much amount of money. For example, in the auction, a developer who bid for 6 million dollars as the second highest valuation, the outsourcer incrementally takes at least 4 million dollars to complete the software.

To solve the problem, we consider the divided ordering method avoiding the risk such as chain-reaction bankruptcy. We assume the dividable module such as assumption 4. Developers bid their valuations with each module like a combinatorial auction [3][1]. Figure 2 shows the example of this situation. In this example, developers 1 serves developing module 1 and 2. Developer 4 has the task of development of module 3 and 4. Here, we give a concrete protocol by using the assumption 4.

Protocol 2

1. For the large-scale software, an outsourcer offers for public subscription. Tasks are divided as multiple modules.
2. Software developers who can contract with the outsourcer come forward as contractor.
3. Developing companies evaluate a value for each module considering the scale of task.
4. Then, they bid their valuations by sealed bid auction. Namely, they bid the set of $\{v_{i1}, \ldots, v_{ij}, \ldots, v_{ik}\}$.
5. The outsourcer calculates a minimized set of all development parties' valuations. Namely, the outsourcer computes $G = \arg\min_i \sum_{j=i}^{k} v_{ij}$.

In this protocol, outsourcer can outsource tasks at the lowest price. The followings are examples of protocol 2.

Example. There are 5 developers. The software consists of 4 modules.
d_1's valuation: $\{v_{11}, v_{12}, v_{13}, v_{14}\}$ is ($\underline{20}$, 60, 40, $\underline{30}$).
d_2's valuation: $\{v_{21}, v_{22}, v_{23}, v_{24}\}$ is (30, $\underline{30}$, 50, 40).
d_3's valuation: $\{v_{31}, v_{32}, v_{33}, v_{34}\}$ is (40, 40, $\underline{20}$, 50).
d_4's valuation: $\{v_{41}, v_{42}, v_{43}, v_{44}\}$ is (25, 50, 50, 70).
d_5's valuation: $\{v_{51}, v_{52}, v_{53}, v_{54}\}$ is (50, 40, 60, 60).

Thus, developer d_1 has implementation of module 1 and 4. Developer d_2 serves the work of module 2. Developer d_3 serves the work of module 3. Total costs of ordering company are calculated as $\sum_{i=1}^{4} v_{ij}$ is 110. We assume that initial fees of the work are paid as fifty percent of the contract prices.

Here, we consider that the developer 3 go out of business due to a certain factor. The outsourcer lost initial fee of developer d_3 for ten dollars. The outsourcer orders and re-allocates the task to the developer d_1 since he/she bids the second lowest value. Totally, the ordering party takes 130 dollars since it needs initial fee of module 3 for developer d_3 and contract fee of module 3 with developer d_1.

Realistically, there are some risks in the protocol 2 since the developer 1 might become bankruptcy. Further, it takes much time to complete all tasks since most of tasks sometimes concentrates to one developer. To solve the problem, we consider the model where each task distributes to more developers.

Figure 3 shows the example of this situation. In this example, each developer serves one task. Thus, time of development is reduced due to distributed task.

Protocol 3

1. For the large-scale software, an outsourcer offers for public subscription. Tasks are divided as multiple modules.
2. Software developers who can contract with the outsourcer come forward as contractor.
3. Developing companies evaluate a value for each module considering the scale of task.

4. Then, they bid their valuations by sealed bid auction. Namely, they bid the set of $\{v_{i1}, \ldots, v_{ij}, \ldots, v_{ik}\}$.
5. The outsourcer calculates a minimized set of all development parties' valuations. Namely, the outsourcer computes $G = \arg\min_i \sum_{j=i}^{k} v_{ij}$ such that each agent serves only one task.

Example. There are 5 developers. The software consists of 4 modules.
d_1's valuation: $\{v_{11}, v_{12}, v_{13}, v_{14}\}$ is (20, 60, 40, $\underline{30}$).
d_2's valuation: $\{v_{21}, v_{22}, v_{23}, v_{24}\}$ is (30, $\underline{30}$, 50, 40).
d_3's valuation: $\{v_{31}, v_{32}, v_{33}, v_{34}\}$ is (40, 40, $\underline{20}$, 50).
d_4's valuation: $\{v_{41}, v_{42}, v_{43}, v_{44}\}$ is ($\underline{25}$, 50, 50, 70).
d_5's valuation: $\{v_{51}, v_{52}, v_{53}, v_{54}\}$ is (50, 40, 60, 60).

In this example, the module m_1 is allocated to the developer d_4. Comparing with the previous example, the allocation of module m_1 changes from developer d_1 to d_3. Thus, the tasks are distributed to avoid non-performance on contract. Here, we give one undesirable example when the all tasks are allocated to one company using protocol 2 and we show the effectiveness of protocol 3.

Example. There are 5 developers. The software consists of 4 modules. The protocol 2 is employed.
d_1's valuation: $\{v_{11}, v_{12}, v_{13}, v_{14}\}$ is ($\underline{50}$, $\underline{40}$, $\underline{50}$, $\underline{60}$).
d_2's valuation: $\{v_{21}, v_{22}, v_{23}, v_{24}\}$ is (90, 60, 70, 70).
d_3's valuation: $\{v_{31}, v_{32}, v_{33}, v_{34}\}$ is (70, 70, 60, 80).
d_4's valuation: $\{v_{41}, v_{42}, v_{43}, v_{44}\}$ is (60, 70, 70, 70).
d_5's valuation: $\{v_{51}, v_{52}, v_{53}, v_{54}\}$ is (80, 50, 60, 70).

In this case using protocol 2, all tasks are allocated to developer d_1 since d_1 bids the lowest valuation to all tasks. However, let us consider the following situation. The conditions of software development company are calculated as $\{A_1, A_2, A_3, A_4, A_5\} = \{0.1, 0.9, 0.9, 0.9, 0.9\}$. Namely, the potential rates of bankruptcy of developers are calculated as $\{r_1, r_2, r_3, r_4, r_5\} = \{0.9, 0.1, 0.1, 0.1, 0.001\}$. If the developer d_1 closes his/her business in period of contract, the outsourcer lost $0.1 \cdot (50+40+50+60) = 20$. Additionally, the tasks are re-allocated to remained developer as follows.

d_2's valuation: $\{v_{21}, v_{22}, v_{23}, v_{24}\}$ is (90, 60, 70, $\underline{70}$).
d_3's valuation: $\{v_{31}, v_{32}, v_{33}, v_{34}\}$ is (70, 70, $\underline{60}$, 80).
d_4's valuation: $\{v_{41}, v_{42}, v_{43}, v_{44}\}$ is ($\underline{60}$, 70, 70, 70).
d_5's valuation: $\{v_{51}, v_{52}, v_{53}, v_{54}\}$ is (80, $\underline{50}$, 60, 70).

Totally, the outsourcer takes 260 (20 + 240). If protocol 3 is employed, the outsourcer takes 245 (5 + 240) even though the developer d_1 becomes bankruptcy.

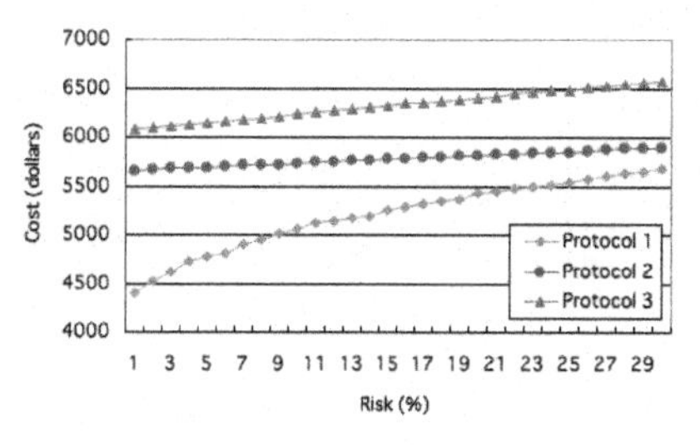

Fig. 4. 3 modules and 5 developers

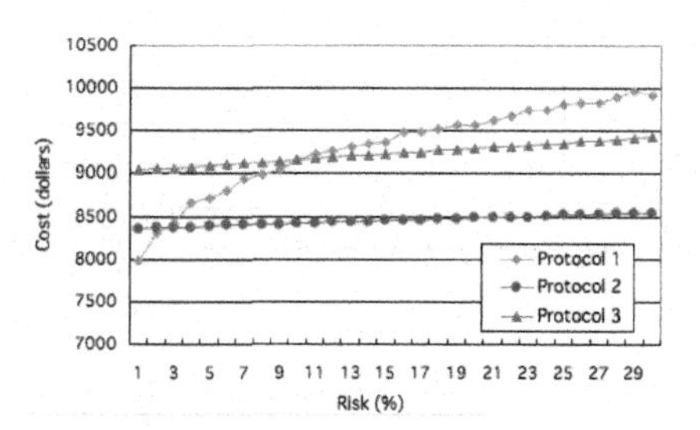

Fig. 5. 5 modules and 8 developers

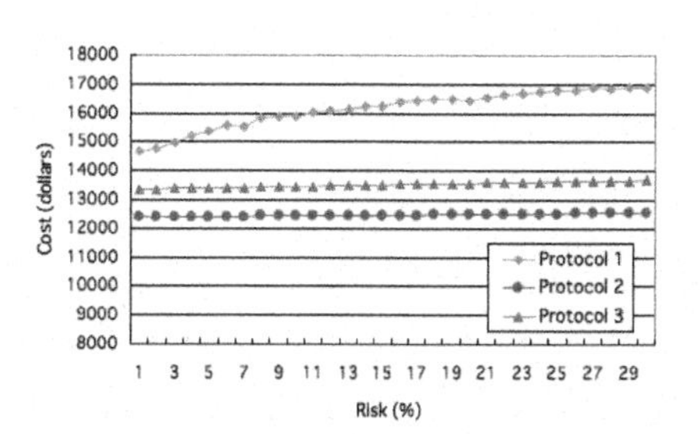

Fig. 6. 8 modules and 12 developers

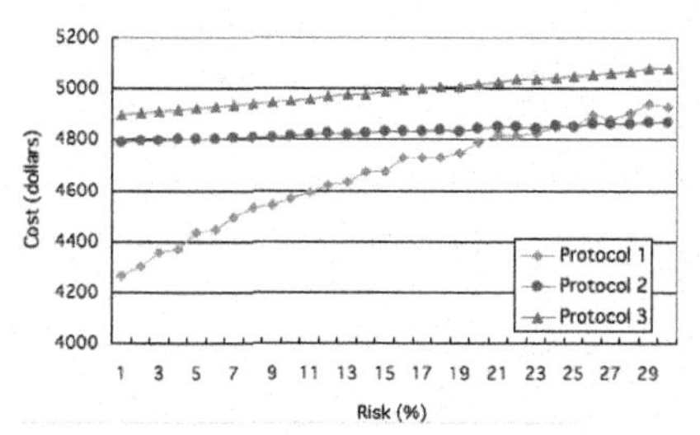

Fig. 7. 3 modules and 10 developers

4 Experiments

To compare and analyze the effectiveness of our proposed issues, we conduct simulations concerned with relationships between rate of risks and outsourcer's cost.

Figures 4 to 9 show experimental results where the number of developers and tasks change. We created 100,000 different problems and show the averages of the cost. The vertical axis shows the average cost for outsourcer. The horizontal axis shows the rate of developer's bankruptcy.

We set the conditions of simulations as follows. Outsourcer orders tasks to developer in prices where developers declare. Namely, the allocations and prices are decided based on sealed first price auction. The cost computation of each task for developers is decided from 1,000 to 4,000 based on uniform distribution. We change the rate of bankruptcy for developers from 1 percent to 30 percent.

Figure 4 shows the result of experiment where the software is divided as 3 modules and 5 developing companies participate in the competition. In this situation, good strategy for outsourcer to reduce cost of ordering is that he/she orders to only one developer. Even though the risks of bankruptcy for developers increase, total cost is less than the protocols 2 and 3.

Figure 5 shows the result of experiment where the software is divided as 5 modules and 8 developing companies participate in the competition. In this situation, good strategy for outsourcer to reduce cost of ordering is that he/she orders to multiple developers distributionally.

Figure 6 shows the result of experiment where the software is divided as 8 modules and 12 developing companies participate in the competition. In this situation,

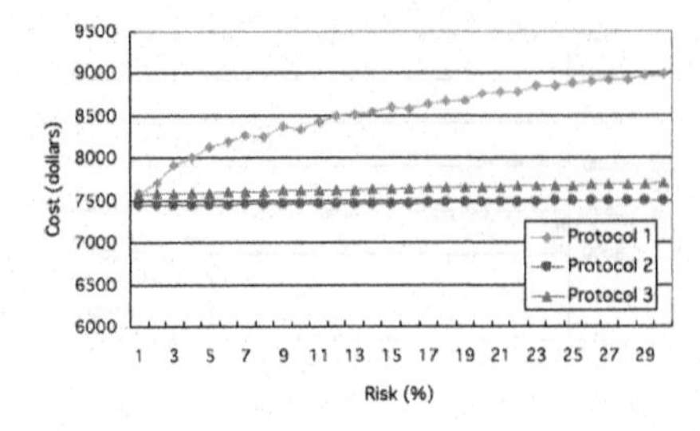

Fig. 8. 5 modules and 16 developers

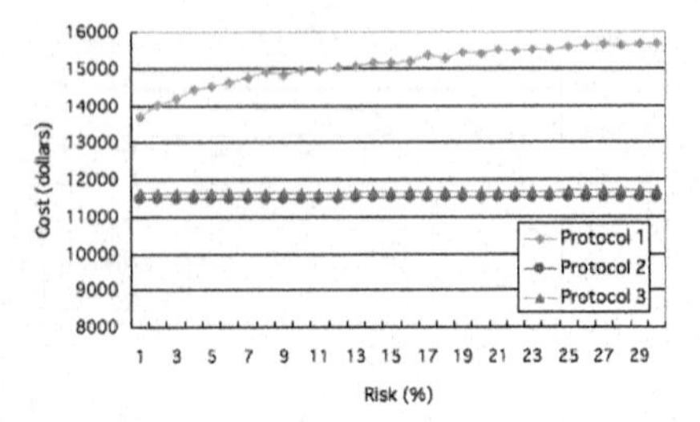

Fig. 9. 8 modules and 24 developers

good strategy for outsourcer to reduce cost of ordering is same as the above second simulation.

Figure 7 shows the result of experiment where the software is divided as 3 modules and 10 developing companies participate in the competition. In this situation, good strategy for outsourcer to reduce cost of ordering is that he/she orders to only one developer. Even though the risks of bankruptcy for developers increase, total cost is less than the protocols 2 and 3. However, we can forecast outsourcer select protocol 2 if the number of developers increases more and more.

Figure 8 shows the result of experiment where the software is divided as 5 modules and 16 developing companies participate in the competition. Figure 9 shows the result of experiment where the software is divided as 8 modules and 24 developers exist. In both situations, the results of simulations are similar. In this situation, good strategy for outsourcer to reduce cost of ordering is that he/she orders to multiple developers distributionally. However, outsourcer may select the protocol 3 since their costs between both protocols are almost same. To employ the protocol 3, the software is completed faster than protocol 2. By just that much, the outsourcer can get much earnings and performance selling/using the software at an early date.

5 Discussion

In the simulation shown at the previous section, task allocations to software developers are decided based on total costs in trading. Allocations are determined based on only valuations bid by developers. In this cases, when a developers gose bankrupt, outsourcer pay a lot of money as the initial payment.

To reduce costs and risks for trading, we propose a protocol where initial payment is determined based on the condition of software development company. We assume the p_{ij}^{pre} is calculated as $p_{ij}^{pre} = r_i \cdot v_{ij}$. Thus, the initial value defined by degree of A_i prevents an outsourcer from intentional bankruptcy by sinister developers.

In our protocol, outsourcer order the module to software developer when the initial payment paid to the developer is low. Even thoug, initial payment is low for the developer, he/she cannot cancel serving tasks on our protocol. We conducted some experiments in this situation.

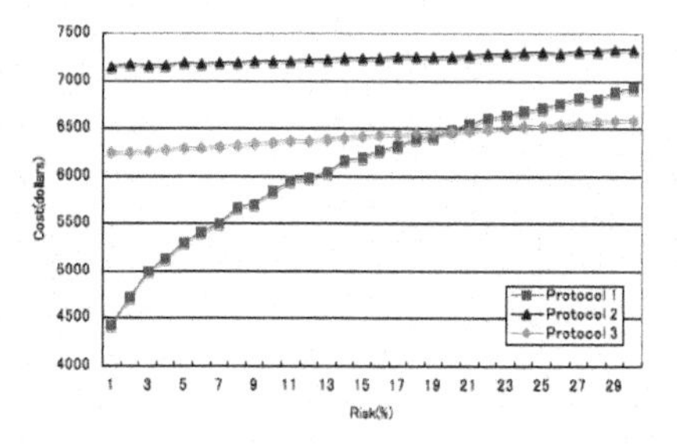

Fig. 10. 3 modules and 5 developers

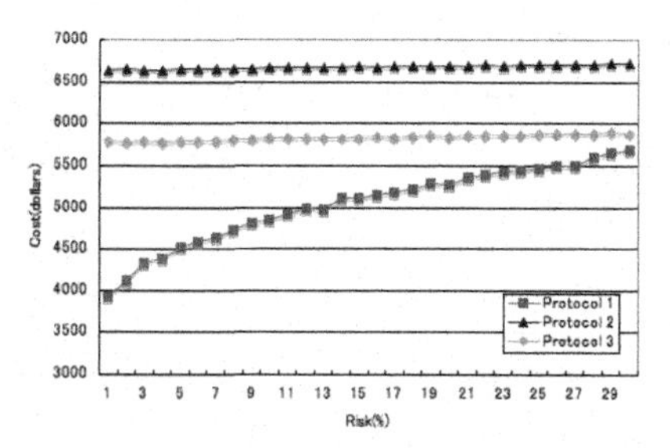

Fig. 11. 3 modules and 10 developers

Example. There are 5 developers. The software consists of 4 modules.
d_1's valuation: $\{v_{11}, v_{12}, v_{13}, v_{14}\}$ is (20, 60, 40, 30).
d_2's valuation: $\{v_{21}, v_{22}, v_{23}, v_{24}\}$ is (40, 30, 50, 40).
d_3's valuation: $\{v_{31}, v_{32}, v_{33}, v_{34}\}$ is (30, 40, 20, 50).
d_4's valuation: $\{v_{41}, v_{42}, v_{43}, v_{44}\}$ is (30, 50, 50, 70).
d_5's valuation: $\{v_{51}, v_{52}, v_{53}, v_{54}\}$ is (50, 40, 60, 60).

d_1's condition: $\{A_1\}$ is (0.7).
d_2's condition: $\{A_2\}$ is (0.6).
d_3's condition: $\{A_3\}$ is (0.5).
d_4's condition: $\{A_4\}$ is (0.5).
d_5's condition: $\{A_5\}$ is (0.4).

d_1's initial payment: $\{p_{11}^{pre}, p_{12}^{pre}, p_{13}^{pre}, p_{14}^{pre}\}$ is (14, 42, 28, 21).
d_2's initial payment: $\{p_{21}^{pre}, p_{22}^{pre}, p_{23}^{pre}, p_{24}^{pre}\}$ is (24, 18, 30, 24).
d_3's initial payment: $\{p_{31}^{pre}, p_{32}^{pre}, p_{33}^{pre}, p_{34}^{pre}\}$ is (15, 20, 10, 25).
d_4's initial payment: $\{p_{41}^{pre}, p_{42}^{pre}, p_{43}^{pre}, p_{44}^{pre}\}$ is (15, 25, 25, 35).
d_5's initial payment: $\{p_{51}^{pre}, p_{52}^{pre}, p_{53}^{pre}, p_{54}^{pre}\}$ is (20, 16, 24, 24).

When outsourcer uses protocol 1 by this example, he/she order all to developer d_3. In this case, initial payment is 70, total costs is 140. If d3 goes bankrupt, outsourcer needs to order all module to another, and needs to pay (70 + new order costs). When outsourcer uses protocol 2 by this example, he/she orders module m_1 and m_4 to developer d_1, he/she orders module m_3 to developer d_3, and he/she orders module m_2 to developer d_5. In this case, initial payment is $\{p_{11}^{pre}, p_{52}^{pre}, p_{33}^{pre}, p_{14}^{pre}\}$ = 14, 16, 10, 21, total costs is 110. If any developer goes bankrupt, the initial payment that he needs to pay is 35 or less. When outsourcer uses protocol 3 by this example, he/she orders module m_1 to developer d_1, he/she orders module m_4 to developer d_2, he/she orders module m_3 to developer d_3, and he/she orders module m_2 to developer d_5. In this case, initial payment is $\{p_{11}^{pre}, p_{52}^{pre}, p_{33}^{pre}, p_{24}^{pre}\}$ = 14, 16, 10, 24, total costs is 120. Total costs is higher than protocol 2. However, initial payment that outsourcer needs to pay is lower than other protocols.

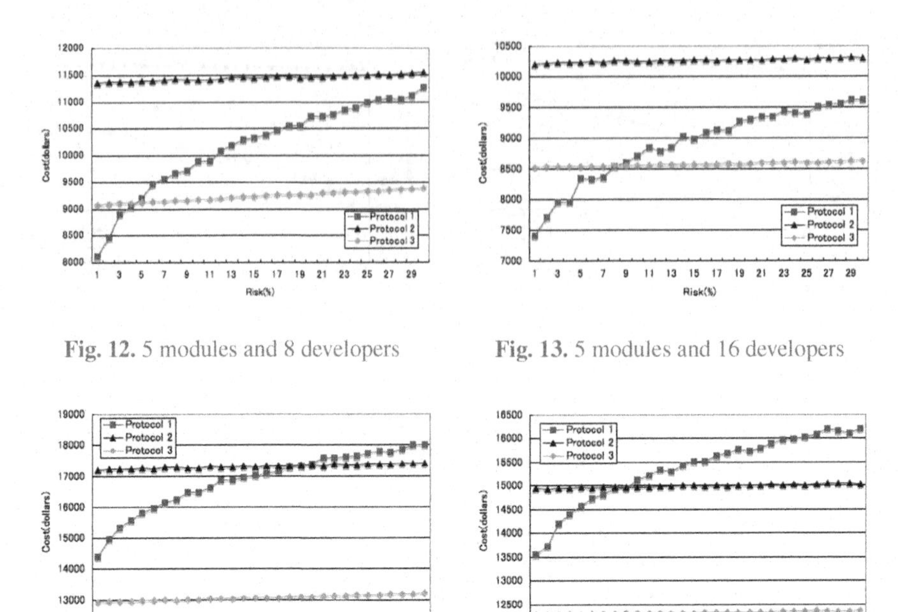

Fig. 12. 5 modules and 8 developers

Fig. 13. 5 modules and 16 developers

Fig. 14. 8 modules and 12 developers

Fig. 15. 8 modules and 24 developers

To compare and analyze the effectiveness of our proposed issues, we conduct simulations on the same condition as Section 4.

Figure 10 shows the result of experiment where the software is divided as 3 modules and 5 developing companies participate in the competition. In this situation, good strategy for outsourcer to reduce cost of ordering is that he/she orders to only one developer. Figure 11 shows the result of experiment where the software is divided as 3 modules and 10 developing companies participate in the competition. In this situation, good strategy for outsourcer to reduce cost of ordering is same as the first simulation. In both situations, even though the risks of bankruptcy for developers increase, total costs less than the protocols 2 and 3.

Figure 12 shows the result of experiment where the software is divided as 5 modules and 8 developing companies participate in the competition. This situation only increases two modules and three companies. However, good strategy changes. Good strategy for outsourcer to reduce cost of ordering is that he/she orders each module to each developer. Figure 13 shows the result of experiment where the software is divided as 5 modules and 16 developing companies participate in the competition. In this situation, good strategy for outsourcer to reduce cost of ordering is same as the third simulation.

Figure 14 shows the result of experiment where the software is divided as 8 modules and 12 developing companies participate in the competition. Figure 15 shows the result of experiment where the software is divided as 8 modules and 24 developing companies participate in the competition. In both situation, good strategy for outsourcer to reduce cost of ordering is that he/she orders each module to each developer. In the situation until now, protocol 1 was cheaper than protocol 2. However, in Figure 14 and Figure 15, protocol 1 is higher than protocol 2. If the outsourcer doesn't know in efficient condition in trading. In such situation, the outsourcer pays a lt of costs when he/she may order all modules to one company.

If all modules are ordered to one company when the number of modules is little, the software is developed at a low cost by the developer. When the number of modules is increased, the costs increase if all modules are ordered to oe software developer. When the large-scale software is divided as many modules, the software manufacture is compleate in shout period and at low price.

6 Conclusions

In this paper, we proposed effective strategies for outsourcer to reduce time and cost with number of modules and developers. This means that the outsourcer should not contract with developers at discretion. Further, in auction to determine subcontractors, outsourcer should gather many bidders. Further, ordering party should divide many modules. However, when the number of modules is less, outsourcer should contract only one developer in figure 4 and 7.

Our future work includes analysis of situation where each scale of modules is different and analysis of situation where integration of modules takes some costs.

References

1. B. Hudson and T. Sandholm, Effectiveness of preference elicitation in combinatorial auctions, Proc. of AAMAS02 Workshop on Agent Mediated Electronic Commerce IV (AMEC IV), 2002.
2. T. Matsuo and Y. Saito, Diversification of risk based on divided tasks in large-scale software system manufacture, Proc. of the 3rd International Workshop on Dataa Engineering Issues in E-Commerce and Services(DEECS 2007), pages 14–25, 2007.
3. D. C. Parkes and L. H. Ungar, Iterative combinatorial auctions: Theory and practice, Proc. of 17th National Conference on Artificial Intelligence (AAAI2000), pages 74–81, 2000.
4. T. Sandholm, Limitations of the vickrey auction in computational multiagent systems, Proceedings of the 2nd International Conference on Multiagent Systems (ICMAS-96), pages 299–306, 1996.
5. T. Sandholm, An algorithm for optimal winnr determination in combinatorial auctions, Proc. of the 16th International Joint Conference on Artificial Intelligence(IJCAI'99), pages 542–547, 1999.
6. P. R. Wurman, M. P. Wellman, and W. E. Walsh, The Michigan internet auctionbot: A configurable auction server for human and software agents, Proc. of the 2nd International Conference on Autonomous Agents (AGENTS98), 1998.

An Approach to Implementing A Threshold Adjusting Mechanism in Very Complex Negotiations: A Preliminary Result

Katsuhide Fujita[1], Takayuki Ito[2], Hiromitsu Hattori[3], and Mark Klein[4]

[1] Department of Computer Science, Nagoya Institute of Technology, Japan.
fujita@longwood.mta.nitech.ac.jp
[2] Techno-Business School / Department of Computer Science
Nagoya Institute of Technology, Japan.
ito.takayuki@nitech.ac.jp
[3] Graduate School of Informatics, Kyoto University, Japan.
hatto@i.kyoto-u.ac.jp
[4] Center for Collective Intelligence, Massachusetts Institute of Technology, USA.
m_klein@mit.edu

Summary. In this paper, we propose a threshold adjusting mechanism in very complex negotiations among software agents. The proposed mechanism can facilitate agents to reach an agreement while keeping their private information as much as possible. Multi-issue negotiation protocols have been studied widely and represent a promising field since most negotiation problems in the real world involve **interdependent multiple issues**. We have proposed negotiation protocols where a bidding-based mechanism is used to find social-welfare maximizing deals. The existing works have not yet concerned about agents' private information. Such private information should be kept as much as possible in their negotiation. Thus, in this paper, we propose a new threshold adjusting mechanism in which agents who open their local information more than the others can persuade the others. The preliminary experimental results demonstrate that the threshold adjusting mechanism can reduce the amount of private information that is required for an agreement among agents.

1 Introduction

Multi-issue negotiation protocols represent an important field of study since negotiation problems in the real world are often complex ones involving multiple issues. While there has been a lot of previous work in this area ([1, 2, 3, 4]) these efforts have, to date, dealt almost exclusively with simple negotiations involving **independent multiple** issues, and therefore linear (single optimum) utility functions. Many real-world negotiation problems, however, involve **interdependent multiple** issues. When designers work together to design a car, for example, the value of a given carburetor is highly dependent on which engine is chosen. The addition of such interdependencies greatly complicates the agent's utility functions, making them nonlinear,

K. Fujita et al.: *An Approach to Implementing A Threshold Adjusting Mechanism in Very Complex Negotiations: A Preliminary Result*, Studies in Computational Intelligence (SCI) **110**, 127–141 (2008)
www.springerlink.com

with multiple optima. Negotiation mechanisms that are well suited for linear utility functions, unfortunately, fare poorly when applied to nonlinear problems ([5]).

We have proposed a bidding-based **multiple-issue** negotiation protocol suited for agents with such nonlinear utility functions. Agents generate bids by sampling their own utility functions to find local optima, and then using constraint-based bids to compactly describe regions that have large utility values for that agent. These techniques make bid generation computationally tractable even in large (*e.g.*, 10^{10} contracts) utility spaces. A mediator then finds a combination of bids that maximizes social welfare.

The existing works have not yet concerned about agents' private information. Such private information should be kept as much as possible in their negotiation. In this paper, we propose a threshold adjusting mechanism. First agents make bids that produce more utility than the common threshold value based on the bidding-based protocol we proposed in [6]. Then the mediator asks each agent to reduce its threshold based on how much each agent opens its private information to the others. Each agent makes bids again above the threshold. This process continues iteratively until agreement is reached or no solution. Our experimental results show that our method substantially outperforms the existing negotiation methods on the point of how much agents have to open their own utility space.

The remainder of the paper is organized as follows. First we describe a model of non-linear multi-issue negotiation. Second, we describe a bidding-based negotiation protocol designed for such contexts. Third we propose a threshold adjusting mechanism that helps agents to keep their private information secret as much as possible. Forth, we present experimental assessment of this protocol. Finally, we compare our work with previous efforts, and conclude with a discussion of possible avenues for future work.

2 Negotiation with Complex Utilities

2.1 Complex Utility Model

We consider the situation where n agents want to reach an agreement. There are m issues, $s_j \in S$, to be negotiated. The number of issues represents the number of dimensions of the utility space. For example, if there are 3 issues[1], the utility space has 3 dimensions. An issue s_j has a value drawn from the domain of integers $[0, X]$, *i.e.*, $s_j \in [0, X]$[2].

[1] The issues are not "distributed" over agents. The agents are all negotiating over a contract that has N (e.g. 10) issues in it. All agents are potentially interested in the values for all N issues.

[2] A discrete domain can come arbitrarily close to a real domain by increasing the domain size. As a practical matter, very many real- world issues that are theoretically real (delivery date, cost) are discretized during negotiations. Our approach, furthermore, is not theoretically limited to discrete domains. The deal determination part is unaffected, though the bid generation step will have to be modified to use a nonlinear optimization algorithm suited to real domains.

A contract is represented by a vector of issue values $\mathbf{s} = (s_1, ..., s_m)$.

An agent's utility function is described in terms of constraints. There are l constraints, $c_k \in C$. Each constraint represents a region with one or more dimensions, and has an associated utility value. A constraint c_k has value $w_i(c_k, \mathbf{s})$ if and only if it is satisfied by contract $\mathbf{s}$. Figure 1 shows an example of a binary constraint between issues 1 and 2. This constraint has a value of 55, and holds if the value for issue 1 is in the range $[3, 7]$ and the value for issue 2 is in the range $[4, 6]$. Every agent has its' own, typically unique, set of constraints.

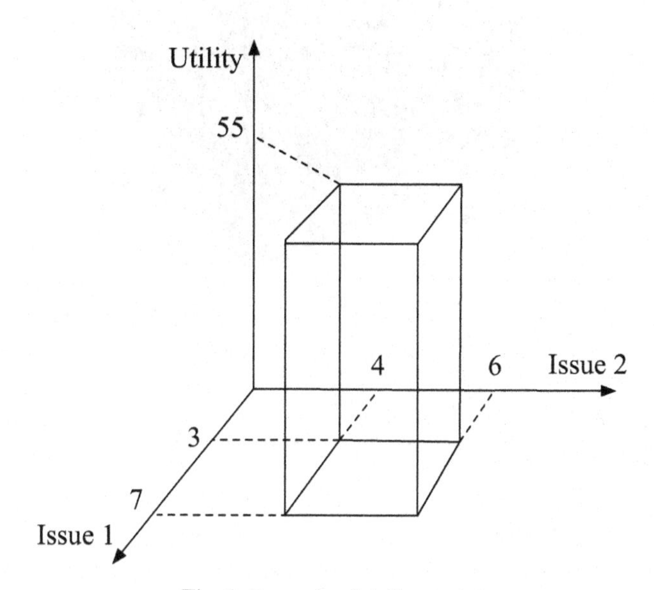

Fig. 1. Example of A Constraint

An agent's utility for a contract s is defined as $u_i(\mathbf{s}) = \sum_{c_k \in C, \mathbf{s} \in x(c_k)} w_i(c_k, \mathbf{s})$, where $x(c_k)$ is a set of possible contracts (solutions) of c_k. This expression produces a "bumpy" nonlinear utility space, with high points where many constraints are satisfied, and lower regions where few or no constraints are satisfied. This represents a crucial departure from previous efforts on multi-issue negotiation, where contract utility is calculated as the weighted sum of the utilities for individual issues, producing utility functions shaped like flat hyper-planes with a single optimum. Figure 2 shows an example of a nonlinear utility space. There are 2 issues, *i.e.*, 2 dimensions, with domains $[0, 99]$. There are 50 unary constraints (*i.e.*, that relate to 1 issue) as well as 100 binary constraints (*i.e.*, that inter-relate 2 issues). The utility space is, as we can see, highly nonlinear, with many hills and valleys.

We assume, as is common in negotiation contexts, which agents do not share their utility functions with each other, in order to preserve a competitive edge. It will generally be the case, in fact, that agents do not fully know their desirable contracts in advance, because each own utility functions are simply too large. If we have 10

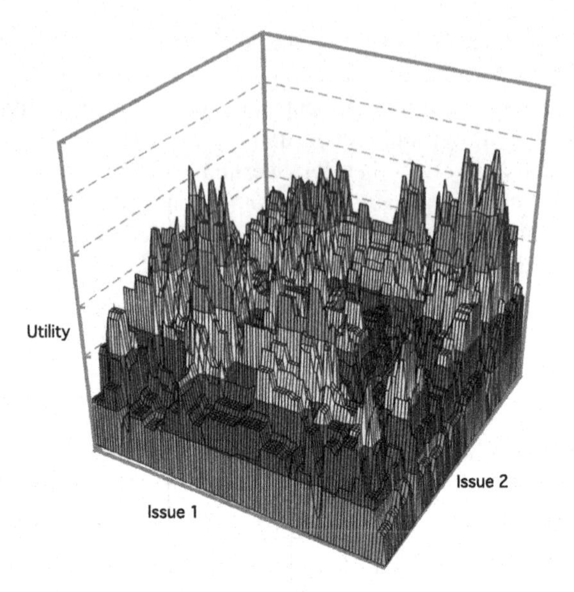

Fig. 2. A Complex Utility Space for a Single Agent

issues with 10 possible values per issue, for example, this produces a space of 10^{10} (10 billion) possible contracts, too many to evaluate exhaustively. Agents must thus operate in a highly uncertain environment.

Finding an optimal contract for individual agents with such utility spaces can be handled using well-known nonlinear optimization techniques such a simulated annealing or evolutionary algorithms. We cannot employ such methods for negotiation purposes, however, because they require that agents fully reveal their utility functions to a third party, which is generally unrealistic in negotiation contexts.

The objective function for our protocol can be described as follows:

$$\arg\max_{\mathbf{s}} \sum_{i \in N} u_i(\mathbf{s}) \quad (1)$$

Our protocol, in other words, tries to find contracts that maximize social welfare, *i.e.*, the total utilities for all agents. Such contracts, by definition, will also be Pareto-optimal.

2.2 Bidding-based Consenting Protocol

The bidding-based negotiation protocol consists of the following four steps:

[Step 1: Sampling] Each agent samples its utility space in order to find high-utility contract regions. A fixed number of samples are taken from a range of random points, drawing from a uniform distribution. Note that, if the number of samples is too low, the agent may miss some high utility regions in its contract space, and thereby potentially end up with a sub-optimal contract.

[Step 2: Adjusting] There is no guarantee, of course, that a given sample will lie on a locally optimal contract. Each agent, therefore, uses a nonlinear optimizer based on simulated annealing to try to find the local optimum in its neighborhood. Figure 3 exemplifies this concept. In this figure, a black dot is a sampling point and a white dot is a locally optimal contract point.

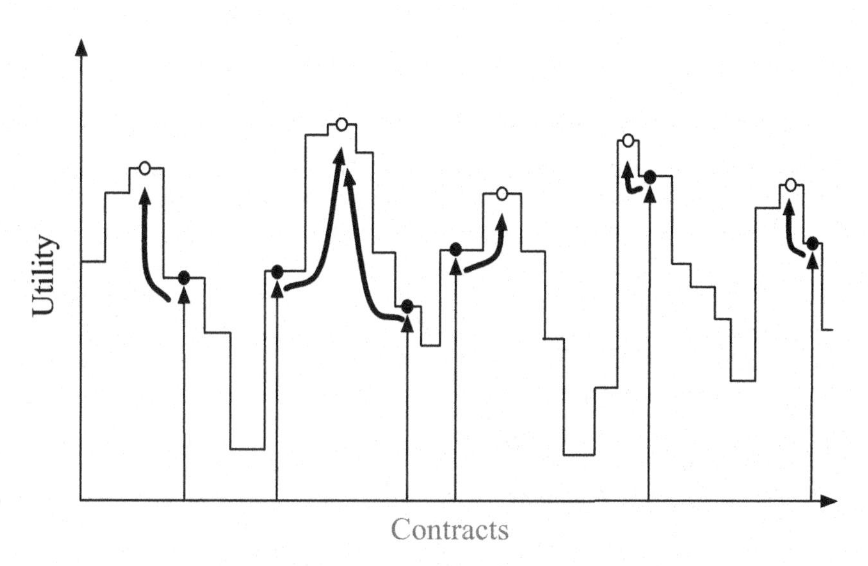

Fig. 3. Adjusting the Sampled Contract Points

[Step 3: Bidding] For each contract s found by adjusted sampling, an agent evaluates its utility by summation of values of satisfied constraints. If that utility is larger than the reservation value δ, then the agent defines a bid that covers all the contracts in the region that has that utility value. This is easy to do: the agent need merely find the intersection of all the constraints satisfied by that s.

Steps 1, 2 and 3 can be captured as follows:
SN: The number of samples
T: Temperature for Simulated Annealing
V: A set of values for each issue, V_m is for an issue m

1: **procedure** bid-generation with SA(*Th*, *SN*, *V*, *T*, *B*)
2: $\quad P_{smpl} := \emptyset$
3: $\quad$ **while** $|P_{smpl}| < SN$
4: $\quad\quad P_{smpl} := P_{smpl} \cup \{p_i\}$ (randomly selected from P)
5: $\quad P := \Pi_{m=0}^{|I|} V_m,\ P_{sa} := \emptyset$
6: $\quad$ **for each** $p \in P_{smpl}$ **do**
7: $\quad\quad p' :=$ simulatedAnnealing(p, T)
8: $\quad\quad P_{sa} := P_{sa} \cup \{p'\}$

9: **for each** $p \in P_{sa}$ **do**
10: $u := 0, B := \emptyset, BC := \emptyset$
11: **for each** $c \in C$ **do**
12: **if** c contains p as a contract and p satisfies c **then**
13: $BC := BC \cup c$
14: $u := u + v_c$
15: **if** $u >= Th$ **then**
16: $B := B \cup (u, BC)$

[Step 4: Deal identification] The mediator identifies the final contract by finding all the combinations of bids, one from each agent, that are mutually consistent, *i.e.*, that specify overlapping contract regions[3]. If there is more than one such overlap, the mediator selects the one with the highest summed bid value (and thus, assuming truthful bidding, the highest social welfare) (see Figure 4). Each bidder pays the value of its winning bid to the mediator.

The mediator employs breadth-first search with branch cutting to find social-welfare-maximizing overlaps:
Ag: A set of agents
B: A set of Bid-set of each agent ($B = \{B_0, B_1, ..., B_n\}$,
A set of bids from agent i is $B_i = \{b_{i,0}, b_{i,1}, ..., b_{i,m}\}$)

1: **procedure** search_solution(B)
2: $SC := \bigcup_{j \in B_0} \{b_{0,j}\}, i := 1$
3: **while** $i < |Ag|$ **do**
4: $SC' := \emptyset$
5: **for each** $s \in SC$ **do**
6: **for each** $b_{i,j} \in B_i$ **do**
7: $s' := s \cup b_{i,j}$
8: **if** s' is consistent **then** $SC' := SC' \cup s'$
9: $SC := SC', i := i + 1$
10: $maxSolution$ = getMaxSolution(SC)
11: **return** $maxSolution$

It is easy to show that, in theory, this approach can be guaranteed to find optimal contracts. If every agent exhaustively samples every contract in its utility space, and has a reservation value of zero, then it will generate bids that represent the agent's complete utility function. The mediator, with the complete utility functions for all agents in hand, can use exhaustive search over all bid combinations to find the social

[3] A bid has an acceptable region. For example, if a bid has a region, such as [0, 2] for issue1, [3,5] for issue2, the bid is accepted by a contract point (1,4), which means issue1 takes 1, issue2 takes 4. If a combination of bids, i.e. a solution, is consistent, there are definitely overlapping region. For instance, a bid with regions (Issue1, Issue2) = ([0, 2], [3, 5]), and another bid with ([0,1], [2,4]) is consistent.

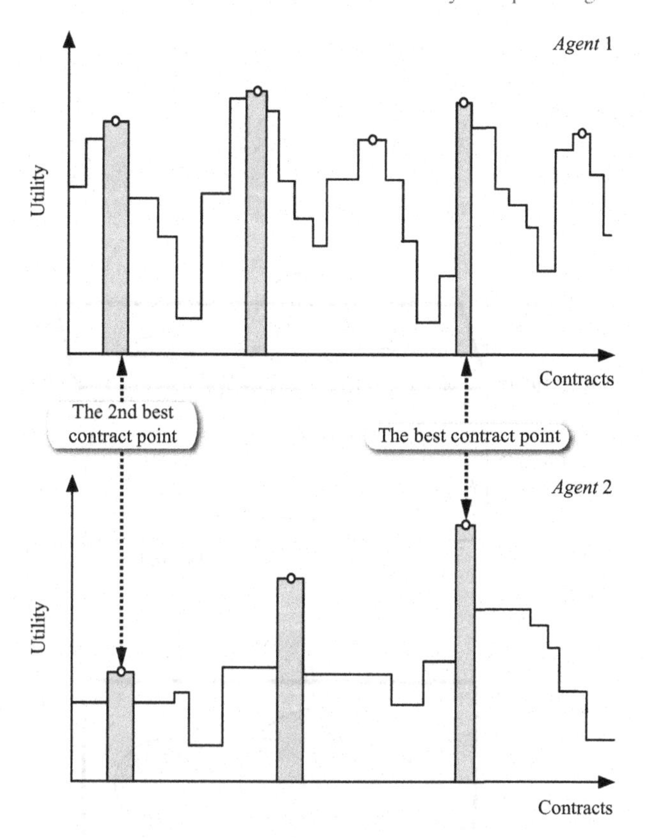

Fig. 4. Deal Identification

welfare maximizing negotiation outcome. But this approach is only practical for very small contract spaces. The computational cost of generating bids and finding winning combinations grows rapidly as the size of the contract space increases. As a practical matter, we introduce the threshold to limit the number of bids the agents can generate. Thus, deal identification can terminate in a reasonable amount of time.

In the previous work [6], the threshold for each agent is commonly defined by the mediator. Agents could not change it by their selves. The threshold adjusting mechanism proposed in this paper allows agents to change their threshold values.

3 Threshold Adjusting Mechanism

3.1 The Outline of the Threshold Adjusting Mechanism

The main idea of the threshold adjusting mechanism is that if an agent reveals the larger area of his utility space, then he can persuade the other agents. On the other hand, an agent who reveals the small area of his utility space, he should adjust his threshold to agree with if no agreement is achieved. The revealed area can be defined how the agent reveals his utility space on his threshold value. The threshold value is defined at the same value beforehand. Then the threshold values are changed by each agent based on the amount of the revealed area afterwards. Figure 5 shows the concept of the revealed area of agent's utility space. If the agent decreases the threshold value, then this means that he reveals his utility space more.

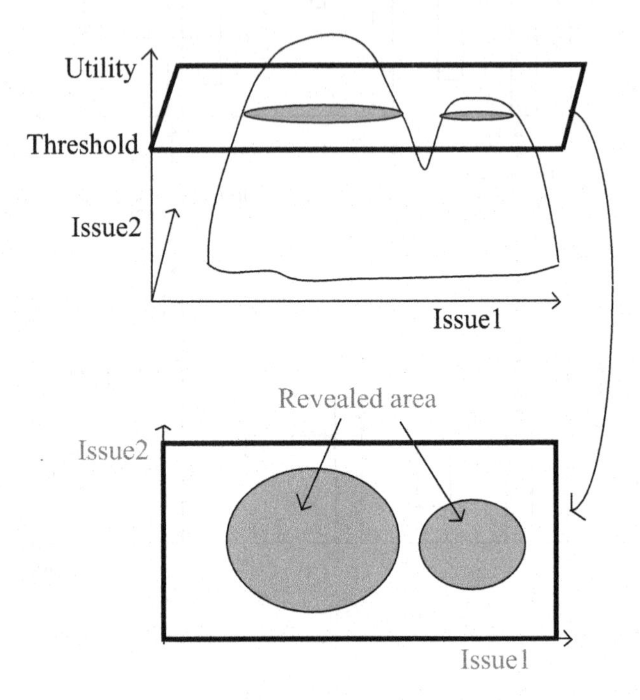

Fig. 5. Revealed Area

Figure 6 shows an example of the threshold adjusting process among 3 agents. The upper figure shows the thresholds and the revealed areas before adjusting the threshold. The bottom figure shows the thresholds and the revealed areas after adjusting the threshold. In particular, in this case, agent 3 revealed the small amount of his utility space. The amount of agent 3's revealing utility space in this threshold adjusting is largest among these 3 agents. In the protocol, this process is repeated until an agreement is achieved or until they could not find any agreement. The exact rate of the amount of revealed utility space and the amount of decreasing the threshold is defined by the mediator or the mechanism designer.

The details of the threshold adjusting mechanism is shown as follows:
Ar: Area of each agent ($Ar = \{Ar_0, Ar_1, ..., Ar_n\}$)

1: **procedure** threshold_adjustment( )

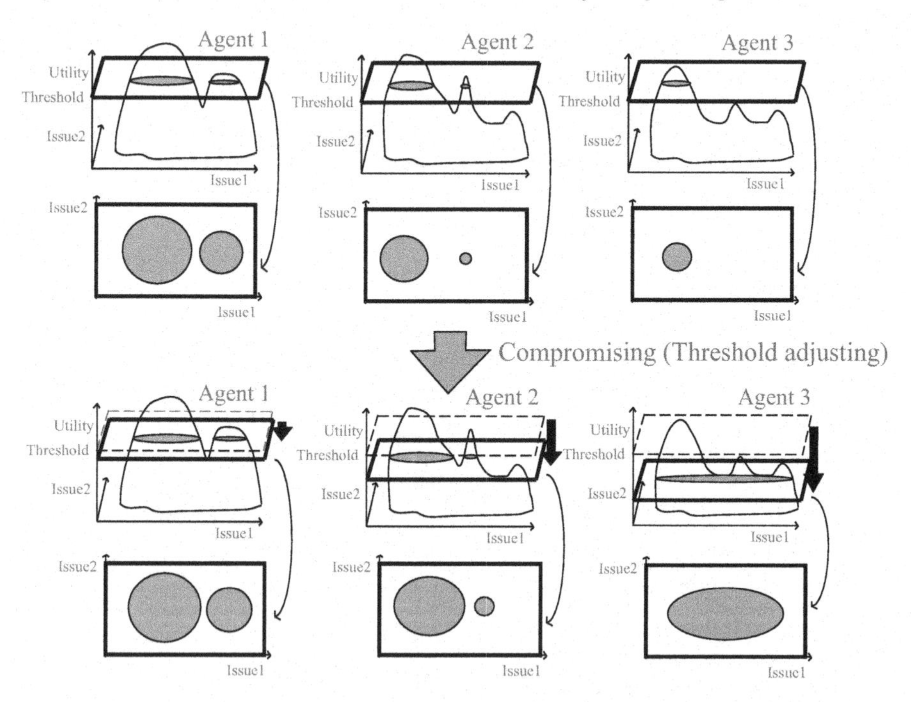

Fig. 6. The Threshold Adjusting Process

```
 2: loop:
 3:   i := 1, B := ∅
 4:   while i < |Ag| do
 5:     bid_generation_with_SA( Th_i, V, SN, T, B_i)
 6:     SC := ∅
 7:     maxSolution := search_solution(B)
 8:   if maxSolution is not empty
 9:     maxSolution := getMaxSolution(SC )
10:     break loop
11:   elseif all agent can lower the threshold
12:     i := 1
13:     SumAr := Σ_{i∈|Ag|} Ar_i
14:     while i < |Ag| do
15:       Th_i := Th_i − C ∗ (σ_i Ar − Ar_i)/σ_i Ar
16:       i := i + 1
17:     end while
18:   else
19:     break loop
```

20: **return** $maxSolution$

The above algorithm utilizes step 1, step 2, step 3, and step 4 in the previous section. In the former paper [6], we did not define any external loop of these steps. This paper is the first that proposed the external loop for an effective consenting mechanism.

3.2 Incremental Deal Identification

The threshold adjusting process shown in the previous section could reduce the computational cost of deal identification in step 4. The original step 4 requires an exponential computational cost because the computation is actually combinatorial optimization. In the new threshold adjusting process, agents incrementally reveal their utility spaces as bids. Thus, for each round, the mediator only computes the new combinations of bids that submitted newly in that round. This process actually reduces the computational cost. We just observed this fact in the preliminary experiments, and did not investigate this deeply. Thus future work includes the investigation of this good feature of incremental deal identification.

4 A Preliminary Experimental Result

We conducted several experiments to evaluate the effectiveness of our approach. In each experiment, we ran 100 negotiations between agents with randomly generated utility functions. We compare our new threshold adjusting protocol and the existing protocol without adjusting the threshold in terms of optimality and privacy.

In the experiments on optimality, for each run, we applied an optimizer to the sum of all the agents' utility functions to find the contract with the highest possible social welfare. This value was used to assess the efficiency (*i.e.*, how closely optimal social welfare was approached) of the negotiation protocols. To find the optimum contract, we used simulated annealing (SA) because exhaustive search became intractable as the number of issues grew too large. The SA initial temperature was 50.0 and decreased linearly to 0 over the course of 2500 iterations. The initial contract for each SA run was randomly selected.

In terms of privacy, the measure is the range of revealed area. Namely, if an agent reveals one point of the gird of utility space, this means he lost 1 privacy unit. If he reveals 1000 points, the he lost 1000 privacy.

The parameters for our experiments were as follows:

Number of agents is $N = 3$. Number of issues is 2 to 10. Domain for issue values is $[0, 9]$.

Constraints : 10 unary constraints, 5 binary constraints, 5 trinary constraints, etc. (a unary constraint relates to one issue, an binary constraint relates to two issues, and so on).

The maximum value for a constraint : $100 \times (Number\ of\ Issues)$. Constraints that satisfy many issues thus have, on average, larger weights. This seems reasonable for many domains. In meeting scheduling, for example, higher order constraints concern more people than lower order constraints, so they are more important for that reason.

The maximum width for a constraint : 7. The following constraints, therefore, would all be valid: issue 1 = $[2, 6]$, issue 3 = $[2, 9]$ and issue 7 = $[1, 3]$.

The number of samples taken during random sampling: $(Number\ of\ Issues) \times$ 200.

Annealing schedule for sample adjustment: initial temperature 30, 30 iterations. Note that it is important that the annealer not run too long or too 'hot', because then each sample will tend to find the global optimum instead of the peak of the optimum nearest the sampling point.

The threshold agents used to select which bids to make in starts with 900 and decreases until 200 in the threshold adjusting mechanism. The protocol without the threshold adjusting process defines the threshold as 200. The threshold is used to cut out contract points that have low utility.

The limitation on the number of bids per agent: $\sqrt[N]{6400000}$ for N agents. It was only practical to run the deal identification algorithm if it explored no more than about 6400,000 bid combinations, which implies a limit of $\sqrt[N]{6400000}$ bids per agent, for N agents.

In our experiments, we ran 100 negotiations in every condition. Our code was implemented in Java 2 (1.5) and run on a core 2 duo processor iMac with 1.0GB memory under Mac OS X 10.4.

Figure 7 shows the optimality of 3 comparable mechanisms, one with the threshold adjustment (with bid limitation), one without the threshold adjustment and bid limitation, and one without the threshold adjustment with bid limitation. The revealed rate is defined by $(Revealedrate) = (Revealedarea)/(Wholeareaofutilityspace)$.

The mechanism without both of the threshold adjustment and bid limitation increasing the revealed rate. This means that if we do not use the threshold adjustment and bid limitation, then agents need to reveal their utility space much more than the other mechanisms.

We found that bid limitation can show the nice effects to keep the increasing amount of revealed rate small. The mechanism with bid limitation but without the threshold adjustment shown by triangles starts decreasing when the number of issue is 5, namely bid limitation starts being active.

Compared with the above two mechanisms, the mechanism with the threshold adjustment proposed inn this paper drastically decreases the amount of the revealed rate.

As we show in the previous paragraph, our proposed threshold adjustment mechanism can effectively reduce the revealed rates. We then show the optimality of our proposed mechanism is quite competitive compared with the other mechanisms in Figure 8.

Each line in Figure 8 means the followings: No threshold adjustment means a mechanism without the threshold adjustment. Threshold adjustment (50), Threshold adjustment (200), and Threshold adjustment (400) mean mechanisms with the threshold adjustment. Each mechanism determines the decreasing amount of the

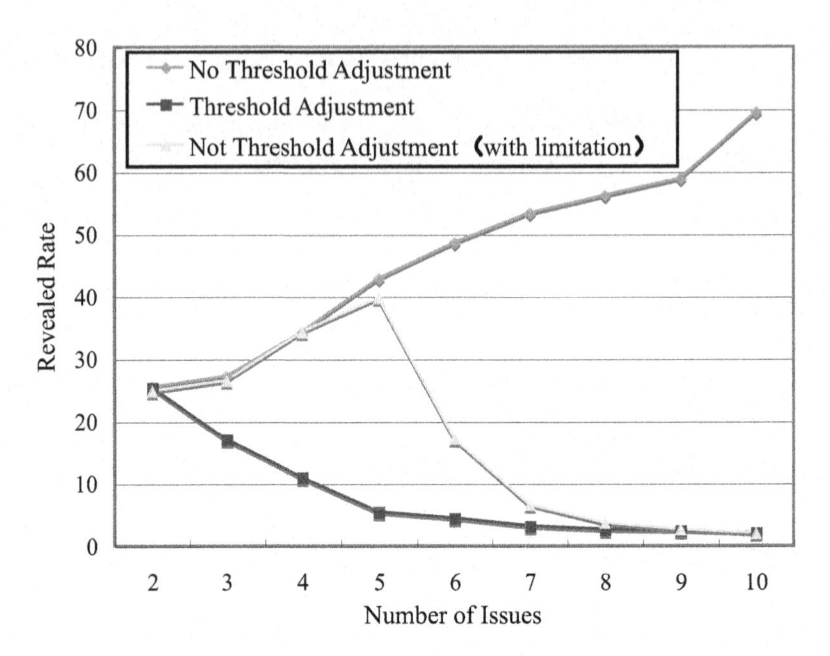

Fig. 7. Revealed Rate

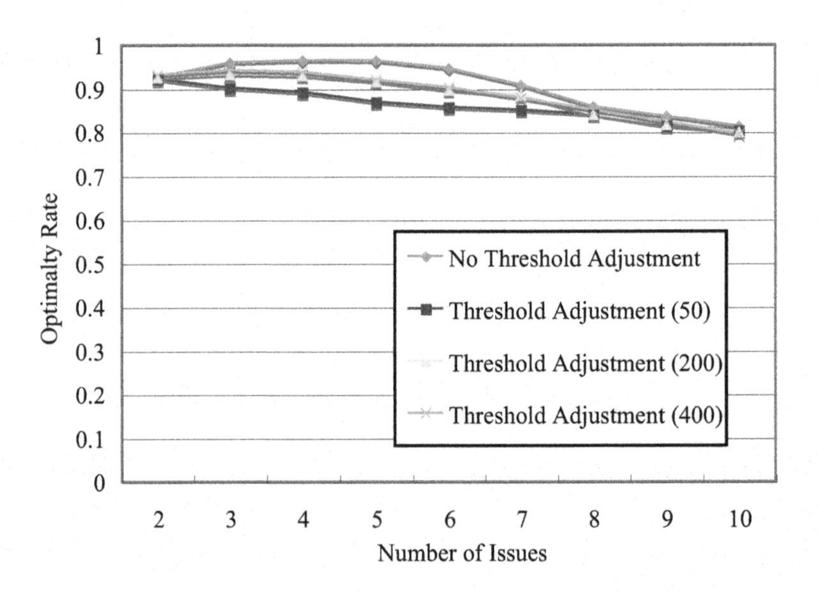

Fig. 8. Optimality

threshold by $50 \times (SumAr - Ar_i)/SumAr$, $200 \times (SumAr - Ar_i)/SumAr$, and $400 \times (SumAr - Ar_i)/SumAr$, respectively. $SumAr$ means the sum of all agents' revealed areas. Ar_i means $agent_i$'s revealed area.

As we can see in Figure 8, in terms of the optimality, the difference between "no threshold adjustment" and "threshold adjustments" is small. At most the difference is around 0.1 around 3 issues to 7 issues. When the threshold decreasing amount is not large, say 50, agents could miss the agreement points that have larger total utilities. This occurs when some agents have higher utility on the agreement point but other agents have very lower utility on the agreement point. "No threshold adjustment" mechanism makes agents to submit all agreement points that have larger utility than the minimum threshold. Thus, "No threshold adjustment" can find such cases. But "threshold adjustment" mechanisms fail to capture such cases when the decreasing amount is smaller.

Figure 9 and Figure 10 compare the required rounds and the revealed rates for different decreasing amounts, 50 and 200. Figure 9 demonstrates that the decreasing amount is small, 50, then the number of rounds could be larger. On the other hand, in Figure 10, if the decreasing amount is small, then the revealed rate is relatively small.

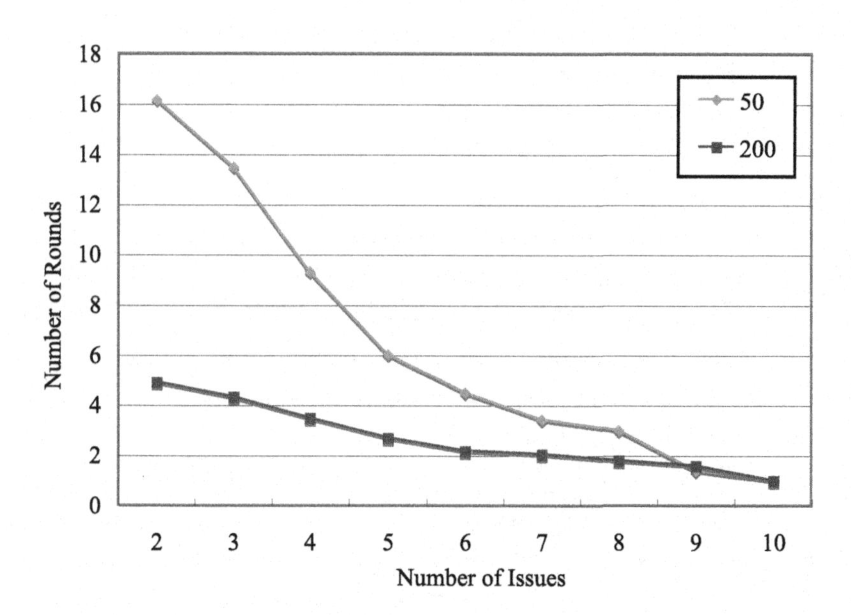

Fig. 9. Number of rounds

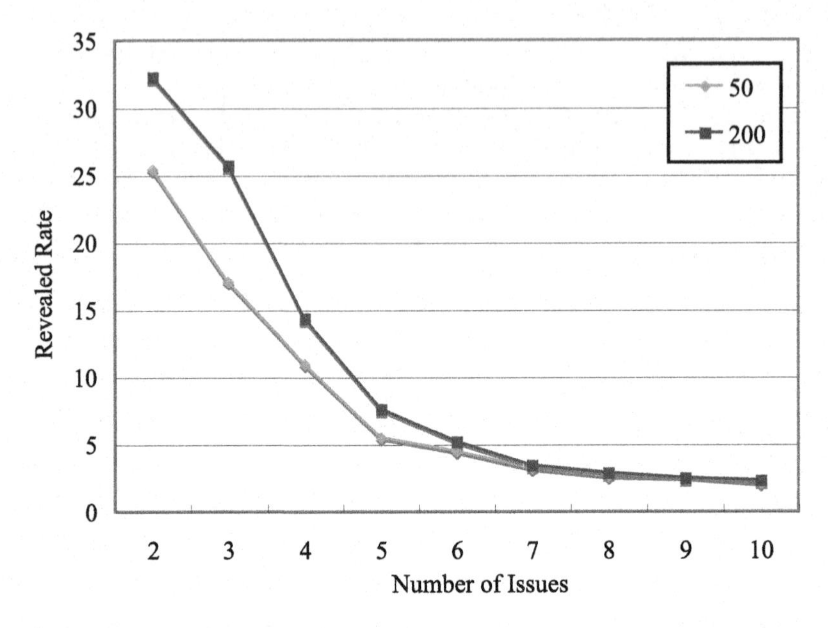

Fig. 10. Revealed Rates

5 Related Work

Most previous work on multi-issue negotiation ([7, 1, 2]) has addressed only linear utilities. A handful of efforts have, however, considered nonlinear utilities. [8] has explored a range of protocols based on mutation and selection on binary contracts. This paper does not describe what kind of utility functions is used, nor does it present any experimental analyses. It is therefore unclear whether this strategy enables sufficient exploration of the utility space to find win-win solutions with multi-optima utility functions. [9] presents an approach based on constraint relaxation. In the proposed approach, a contract is defined as a goal tree, with a set of on/off labels for each goal, which represents the desire that an attribute value is within a given range. There are constraints that describe what patterns of on/off labels are allowable. This approach may face serious scalability limitations. However, there is no experimental analysis and this paper presents only a small toy problem with 27 contracts. [10] also presents constraint based approach. In this paper, a negotiation problem is modeled as a distributed constraint optimization problem. During exchanging proposals, agents relax their constraints, which express preferences over multiple attributes, over time to reach an agreement. This paper claims the proposed algorithm is optimal, but do not discuss computational complexity and provides only a single small-scale example. [5] presented a protocol, based on a simulated-annealing mediator, that was applied with near-optimal results to medium-sized bilateral negotiations with binary dependencies. The work presented here is distinguished by demonstrating both scalability, and high optimality values, for multilateral negotiations and higher order dependencies.

6 Conclusions

In this paper, we proposed a threshold adjusting mechanism in very complex negotiations among software agents. In very complex negotiations, we assume agents have to do interdependent multi-issue negotiation. The threshold adjusting mechanism can facilitate agents to reach an agreement while keeping their private information as much as possible. The preliminary experimental results demonstrate that the threshold adjusting mechanism can reduce the amount of private information that is required for an agreement among agents. One of the interesting future works includes more autonomous threshold adjustment mechanism for agents in very complex negotiations.

References

1. P. Faratin, C. Sierra, and N.R. Jenning. Using similarity criteria to make issue trade-offs in automated negotiations. In *Artificial Intelligence*, pp. 142:205–237, 2002.
2. S. Fatima, M. Wooldridge, and N.R. Jennings. Optimal negotiation of multiple issues in incomplete information settings. In *AAMAS-2004*, pp. 1080–1087, 2004.
3. R. Y. K. Lau. Towards genetically optimised multi-agent multi-issue negotiations. In *of HICSS-2005*, 2005.
4. L.K. Soh and X. Li. Adaptive, confidence-based multiagent negotiation strategy. In *AAMAS-2004*, pp. 1048–1055, 2004.
5. M. Klein, P. Faratin, H. Sayama, and Y. Bar-Yam. Negotiating complex contracts. *Group Decision and Negotiation*, 12(2):58–73, 2003.
6. T. Ito, H. Hattori, and M. Klein. Multi-issue negotiation protocol for agents : Exploring nonlinear utility spaces. In *JCAI-2007*, pp. 1347–1352, 2007.
7. T. Bosse and C. M. Jonker. Human vs. computer behaviour in multi-issue negotiation. In *RRS-2005*, pp. 11–24, 2005.
8. R. J. Lin and S. T. Chou. Bilateral multi-issue negotiations in a dynamic environment. In *AMEC-2003*, 2003.
9. M. Barbuceanu and W. K. Lo. Multi-attribute utility theoretic negotiation for electronic commerce. In *AMEC-2000*, pp. 15–30, 2000.
10. X. Luo, N. R. Jennings, N. Shadbolt, H. Leung, and J.H. Lee. A fuzzy constraint based model for bilateral, multi-issue negotiations in semi-competitive environments. *Artificial Intelligence*, 148:53–102, 2003.

A New Approach to Detecting False-Name Bids in e-Auctions

Tokuro Matsuo*, Takayuki Ito and Toramatsu Shintani

Faculty of Engineering,
Yamagata University,
4-3-16, Jonan, Yonezawa, Yamagata, 992-8510, Japan.
E-mail: matsuo@yz.yamagata-u.ac.jp
*Corresponding author

Summary. This paper presents a method for discovering and detecting shill bids in combinatorial auctions. Combinatorial auctions have been studied very widely. The Generalized Vickrey Auction (GVA) is one of the most important combinatorial auctions because it can satisfy the strategy-proof property and Pareto efficiency. As some literatures pointed out, false-name bids and shill bids pose an emerging problem for auctions, since on the Internet it is easy to establish different e-mail addresses and accounts for auction sites. GVA cannot satisfy the false-name-proof property. Moreover, they proved that there is no auction protocol that can satisfy all three of the above major properties. Their approach concentrates on designing new mechanisms. As a new approach against shill-bids, in this paper, we propose a method for finding shill bids with the GVA in order to avoid them. Our algorithm can judge whether there might be a shill bid from the results of the GVA's procedure. However, a straightforward way to detect shill bids requires an exponential amount of computing power because we need to check all possible combinations of bidders. Therefore, in this paper we propose an improved method for finding a shill bidder. The method is based on winning bidders, which can dramatically reduce the computational cost. The results demonstrate that the proposed method successfully reduces the computational cost needed to find shill bids. The contribution of our work is in the integration of the theory and detecting fraud in combinatorial auctions.

1 Introduction

This paper presents a method for detecting shill bids in combinatorial auctions. The purpose of it is an important issue not to reduce sellers' revenues due to shills and frauds. Auction theory has received much attention from computer scientists and economic scientists in recent years. One reason for this interest is the fact that Internet auctions such as Yahoo Auction and eBay have developed very quickly and widely. Also, auctions in B2B trades are increasing rapidly. Moreover, there is growing interest in using auction mechanisms to solve distributed resource allocation problems in the field of AI and multi-agent systems.

Combinatorial auctions have been studied very widely as one of most important auction formats. In a combinatorial auction, bidders can make bids on bundles of

T. Matsuo et al.: *A New Approach to Detecting False-Name Bids in e-Auctions*, Studies in Computational Intelligence (SCI) **110**, 143–156 (2008)
www.springerlink.com

multiple different items. The main issue in a combinatorial auction is its winner determination problem. The computation for winner determination is an NP-complete problem, since there can be exponential numbers of bundles of items and an auctioneer needs to find a bundle combination that maximizes revenue. Many studies have approached this problem by investigating a variety of search algorithms.

The Generalized Vickrey Auction (GVA) is one of the combinatorial auctions that are strategy-proof, i.e., the dominant strategy is for bidders to declare their true evaluation value for a good, and its allocation is Pareto efficient. Many scientists in the field of auction theory have focused on GVA because of its strategy-proof property. The details of GVA are described in Section 2.

As some literatures pointed out, false-name bids and shill bids are an emerging problem for auctions, since on the Internet it is easy to establish different e-mail addresses and accounts for auction sites. Bidders who make false-name bids and shill bids can benefit if the auction is not robust against false-name bids. A method to avoid false-name bids and shill bids is a major issue that auction theorists need to resolve.

GVA cannot satisfy the false-name-proof property. Moreover, it is proved that there is no auction protocol that can satisfy all of the three major properties, i.e., false-name-proof property, Pareto efficiency, and strategy-proof property. Thus, some existing researches have developed several other auction protocols that can satisfy at least the false-name-proof property. However, no existing researches can detect shill bidders in the process of their auctions. In this paper, we give and discuss the way how shills are detected. Furthermore, we give features and characteristics about shill bidding in combinatorial auctions for our work. In actual auctions, it is an important issue in which auctioneers and sellers can know who is shill and fraud bidders. They can use such useful information in subsequent auctions and trades. Concretely, in this paper we propose a method for finding shill bids with GVA in order to avoid them. Our algorithm can judge whether there might be a shill bid from the results of the GVA's procedure. If there is the possibility of a shill bid, the auctioneer can make the decision of whether to stop allocation of items based on the results. Namely, in our approach, we build an algorithm to find shill bids in order to avoid them. This differs from the approach of Yokoo et al., which builds mechanisms to avoid shill bids.

A shill bid is defined as two or more bids created by a single person, who can unfairly gain a benefit from creating such bids. Therefore, the straightforward way to find shill bids is to find a bidder whose utility becomes negative when his/her bids and those of another bidder are merged. The merging method is described in Section 3. However, this straightforward method requires an exponential amount of computing power, since we need to check all of combinations of bidders.

Thus, in this paper, we propose an improved method for finding a shill bidder. The method is based on the brute force algorithm, and it can dramatically reduce the computational cost.

In this paper, we concentrate the detecting method shill bidders. If the auctioneer can know that shill bidders include in all bidders in an auction, the auctioneer can cease the auction to avoid reducing sellers' earnings. However, using the information detected shills, new auction mechanism can be invented, that is, the system allocates items on expensive prices to the shills. In such auction mechanism, no shill bidders have incentives scheming shills.

The rest of this paper consists of the following six parts. In Section 2, we show preliminaries on several terms and concepts of auctions. In Section 3, shill-biddable allocations are defined and discussed. In Section 4, the brute force algorithm used in this paper is introduced. In Section 5, we explain how we handle a massive number of bidders. Finally, we present our concluding remarks and future work.

2 Related Work

Milgrom analyzed the shill-biddable feature in VCG [6]. Bidders in GVA can profitably use shill bidders, intentionally increasing competition in order to generate a lower price. Thus, the Vickrey auction provides opportunities and incentives for collusion among the low-value, losing bidders. This feature is a result of the monotonic increase problem. However, this work does not refer to the method of detecting shill bidding in combinatorial auctions.

Yokoo et al. reported the effect of false-name bids in combinatorial auctions [12]. To solve the problem, Yokoo, Sakurai and Matsubara proposed novel auction protocols that are robust against false-name bids [10]. The protocol is called the Leveled Division Set (LDS) protocol, which is a modification of the GVA and it utilizes reservation prices of auctioned goods for making decisions on whether to sell goods in a bundle or separately. Furthermore, they also proposed an erative Reducing(IR) protocol that is robust against false-name bids in multi-unit auctions [11]. The IR protocol is easier to use than the LDS, since the combination of bundles is automatically determined in a flexible manner according to the declared evaluation values of agents. They concentrate on designing mechanisms that can be an alternative of GVA. On the other hand, detecting shills is important issues for auctioneers to make effective auction markets. Due to our fundamentally different purpose, we do not simply adopt off-the-shelf methods for mechanism design.

Some researchers [7][8][5][4][2] proposed methods for computing and calculating the optimal solution in combinatorial auctions. These analyses contributed to the pursuit of a computational algorithm for winner determination in combinatorial auctions, but they did not deal with shill bidding and thus are fundamentally different approaches from our work. However, some of these algorithms can be incorporated in our work. Combinatorial auctions have a computationally hard problem in which the number of combinations increases when the number of participants/items increases in an auction, since agents can bid their evaluation values as a set of bundled items.

Sandholm [7] propose a fast winner determination algorithm for combinatorial auctions. Also, Sandholm [8] showed how different features of a combinatorial market affect the complexity of determining the winners. They studied auctions, reverse auctions, and exchanges, with one or multiple units of each item, with and without free disposal. We theoretically analyzed the complexity of finding a feasible, approximate, or optimal solution.

Fujishima et al. proposed two algorithms to mitigate the computational complexity of combinatorial auctions [2]. Their proposed Combinatorial Auction Structured Search (CASS) algorithm determines optimal allocations very quickly and also provides good "any-time" performance. Their second algorithm, called VSA, is based on a simulation technique. CASS considers fewer partial allocations than the brute force method because it structures the search space to avoid considering allocations containing conflicting bids. It also caches the results of partial searches and prunes the search tree. On the other hand, their second algorithm, called Virtual Simultaneous Auction (VSA), generates a virtual simultaneous auction from the bids submitted in a real combinatorial auction and then carries out simulation in the virtual auction to find a good allocation of goods for the real auction. In our work, to determine optimal allocations quickly in each GVA, we employ the CASS method. However, Fujishima's paper does not focus on shill bids.

Leyton-Brown et al. proposed an algorithm for computing the optimal winning bids in a multiple units combinatorial auction [5]. This paper describes the general problem in which each good may have multiple units and each bid specifies an unrestricted number of units desired for each good. The paper proves the correctness of our branch-and-bound algorithm based on a dynamic programming procedure. Lehmann et al. proposed a particular greedy optimization method for computing solutions of combinatorial auctions [4]. The GVA payment scheme does not provide for a truth-revealing mechanism. Therefore, they introduced another scheme that guarantees truthfulness for a restricted class of players.

3 Preliminaries

3.1 Model

Here, we describe a model and definitions needed for our work. The participants of trading consist of a manager and bidders. The manager prepares multiple items, and bidders bid evaluation values for what they want to purchase.

- In an auction, we define that a set of bidders/agents is $N = \{1, 2, \ldots, i, \ldots, n\}$ and a set of items is $G = \{a_1, a_2, \ldots, a_k, \ldots, a_m\}$.
- $v_i^{a_k}$ is bidder i's evaluation value at which the ith bidder bids for the kth item $(1 \leq i \leq n, 1 \leq k \leq m)$.
- $v_i(B_i^{a_k,a_l})$ is bidder i's evaluation value at which the ith bidder bids for the bundle including the kth and lth items $(1 \leq i \leq n, 1 \leq k, l \leq m)$. The form of this description is used when the bidder evaluates more than two items.
- $p_i^{a_k}$ is the payment when agent i can purchase an item a_k. When the bidder i purchases the set of bundles of items, the payment is shown as $p_i(B_i^{a_k,a_l})$.
- The set of choices is $G = \{(G_1, \ldots, G_n) : G_i \cap G_j = \phi, G_i \subseteq G\}$.
- G_i is an allocation of a bundle of items to agent i.

Assumption 1 (Quasi-linear utility) *Agent i's utility u_i is defined as the difference between the evaluation value v_i of the allocated good and the monetary transfer p_i for the allocated good. $u_i = v_i - p_i$. Such a utility is called a quasi-linear utility, and we assume the quasi-linear utility.*

Assumption 2 (Monotonicity of evaluation values) *Regarding the true evaluation values of any bidder, for bundles B and B', if $B \subset B'$, $B \neq B'$, $v_i(B,\theta_i) \leq v_i(B',\theta_i)$ holds.*

Namely, in this paper, when the number of items in a bundle increases, the total evaluation values for the bundle decrease. This means that free disposal is assumed.

Assumption 3 (Only one bidder engaging in shill bids) *For simplicity, in this paper, we assume that only one schemer is engaging in shill bids.*

The reason why we set the above assumption is as follows. First, in combinatorial auctions, the solution spaces are sometimes increased exponentially as huge spaces. When the numbers of items and bidding are increased, it takes a lot of minutes to find the optimal solutions/allocations. Second, in the combinatorial auction research, the analyses of determination of allocations, detection of shills and calculation of payment amount are very complicated problems. The method of detecting shills in multiple schemers is almost same as a case of one schemer. However, when a schemer bids the bid values with non single-minded/tie break, the inspection accuracy reduces because the payment amount changes due to the allocation change.

Each bidder i has preferences for the subset $G_i \subseteq G$ of goods, which here is considered a bundle. Formally, each bidder has type θ_i, that is, in a type's set Θ. Based on the type, we show that the bidder's utility is $v_i(G_i, \theta_i) - p_i^{G_i}$ when the bidder purchases item G_i for $p_i^{G_i}$. Note that $(v_i(G_i, \theta_i))$ is bidder i's evaluation value of bundle $G_i \subseteq G$.

3.2 GVA: Generalized Vickrey Auction

GVA was developed from the Vickrey-Clarke-Groves mechanism [9] [1][3] which is strategy-proof and Pareto efficient if there exists no false-name bid. We say an auction protocol is Pareto efficient when the sum of all participants' utilities (including that of the auctioneer), i.e., the social surplus, is maximized. If the number of goods is one, in a Pareto efficient allocation, the good is awarded to the bidder having the highest evaluation value corresponding the quality of the good.

In the GVA, first each agent tells his/her evaluation value $v_i(G_i, \theta_i)$ to the seller. We omit the "type" notation and simply write $v_i(G_i, \theta_i) = v_i(G_i)$. The efficient allocation is calculated as an allocation to maximize the total value:

$$G^* = arg \max_{G=(G_1,\ldots,G_n)} \sum_{i \in N} v_i(G_i).$$

The auctioneer informs the payment amount to the bidders. Agent i's payment p_i is defined as follows.

$$p_i = \sum_{i \neq j} v_j(G^*_{\sim i}) - \sum_{i \neq j} v_j(G^*).$$

Here, $G^*_{\sim i}$ is the allocation that maximizes the sum of all agents' evaluation values other than agent i's value. Except for agent i, it is the allocation in which the total evaluation value is maximum when all agents bid their evaluation values:

$$G^*_{\sim i} = \arg\max_{G \backslash G_i} \sum_{N-i} v_j(G_j).$$

3.3 Example of shill bids

In auction research, some papers have reported the influence of false name bids in combinatorial auctions, such as GVA [12]. These are called "shill bids." Here, we show an example of shill bids.

Assume there are two bidders and two items. Each agent bids for a bundle, that is, $\{a_1, a_2, (a_1, a_2)\}$.

Agent 1's evaluation value $v_1(B_1^{a_1,a_2}) : \{\$6, \$6, \$12\}$
Agent 2's evaluation value $v_2(B_2^{a_1,a_2}) : \{\$0, \$0, \$8\}$

In this case, both items are allocated to agent 1 for 8 dollars. Agent 1's utility is calculated as $12 - 8 = 4$.

If agent 1 creates a false agent 3, his/her utility increases.

Agent 1's evaluation value $v_1(B_1^{a_1,a_2}) : \{\$6, \$0, \$6\}$
Agent 2's evaluation value $v_2(B_2^{a_1,a_2}) : \{\$0, \$0, \$8\}$
Agent 3's evaluation value $v_3(B_3^{a_1,a_2}) : \{\$0, \$6, \$6\}$

Agent 1 can purchase item a_1 and agent 3 can purchase item a_2. Each agent's payment amount is $8 - 6 = 2$ and each agent's utility is calculated as $6 - 2 = 4$. Namely, agent 1's utility is 8 dollars (because agent 3 is the same as agent 1).

3.4 Impossibility Theorem

Yokoo et al. examined the effect of false-name bids on combinatorial auction protocols [12]. False-name bids are bids submitted by a single bidder using multiple identifiers such as multiple e-mail addresses. They showed a formal model of combinatorial auction protocols in which false-name bids are possible. The obtained results are summarized as follows: (1) the Vickrey-Clarke-Groves (VCG) mechanism, which is strategy-proof and Pareto efficient when there exists no false-name bid, is not falsename-proof; (2) there exists no false-name-proof combinatorial auction protocol that satisfies Pareto efficiency; (3) one sufficient condition where the VCG mechanism is false-name-proof is identified, i.e., the concavity of a surplus function over bidders.

4 Shill-biddable Allocation

4.1 Definition

We define a shill-biddable allocation as an allocation where an agent can increase his/her utility by creating shill bidders The word "shill bid" has generally multiple meanings. In our paper, "shill bid" is used as same meaning with false-name bid, that is a sheme agent produces false-agents.[1] Some early works show and discuss concerned with false-name bidding, which a bidder agent makes false-agents and he/she can increase the utility producing the shill bidders. In our work, we use a term of the "shill bidding", that is an extended conception of false-name bid. Then, how can we judge and know that the allocation is shill-biddable In general, shill bidders are produced by a bidder who is up to increasing his/her utility. We propose a method for judging and discovering allocations that may be susceptible to shill bids ("shill-biddable allocation"). When shill bidders are created, the number of agents increases in the situation where no shill bidder is created. When the number of real bidders is n and the total number of bids is $n + n'$, we find that the number of shill bidders is n'. However, we can know what kind auctions have shill biddable possibility.

Definition 1 (Shill-biddable allocation). *A shill biddable allocation is an allocation where agents who create shill bidders can increase their utility over that of an agent who does not create shill bidders in an auction.*

When agents create shill bidders in the auction, the agents' utilities increase. For example, we assume that winner agent i's utility is $u_i(B_i^{a_1,\dots,a_k,\dots,a_m})$ in an auction. $u_i(B_i^{a_1,\dots,a_m})$ is agent i's total utilities including a set of bundled items $(a_1,\dots,a_m)$, where agent i does not create shill bidders in the auction.

Here, we consider the situation where the agent creates shill bidder u'_i. The original agent bids a set of bundled items $(a_1,\dots,a_k)$ and the shill agent bids another set of bundled items $(a_{k+1},\dots,a_m)$. We assume that each agent can purchase items for which he/she bids. The original agent's utility is $u_i(B_i^{a_1,\dots,a_k})$ and the shill agent's utility is $u'_i(B_i^{a_{k+1},\dots,a_m})$, where agent i makes a shill bidder in the auction.

A shill-biddable allocation is defined when the following equation holds.

$$u_i(B_i^{a_1,\dots,a_k,\dots,a_m}) \le F(u_i(B_i^{a_1,\dots,a_k}), u'_i(B_i^{a_{k+1},\dots,a_m})).$$

In the above equation, $F(\cdot)$ means a kind of merge function, e.g. summation or maximization, etc. In this paper, we assume that the merge function is a maximizing function because maximum values are the evaluation values at which the agent can pay for the item.

Under the **Assumption 1**, we can show the concrete merge equation as follows.

$$\begin{aligned} & F(u_i(B_i^{a_1,\dots,a_k}), u'_i(B_i^{a_{k+1},\dots,a_m})) \\ = & \{\max\{u_i(B_i^{a_1,\dots,a_k,\dots,a_m}), u'_i(B_i^{a_1,\dots,a_k,\dots,a_m})\}\} \end{aligned}$$

[1] In general, "shill bidding" means decoy bids among participants that include sellers. In this paper, we use the term "shill bidding" that means decoy bids among only bidders in the same sense as the term "false-name bidding."

For example, we consider a merge investment by using the following two agents' values. Agent 1's value $v_1(B_1^{a_1,a_2,a_3})$ is $\{2, 6, 5, 8, 7, 11, 13\}$ and agent 2's value $v_2(B_2^{a_1,a_2,a_3})$ is $\{5, 4, 6, 7, 10, 9, 15\}$. In this case, merge investment $v_{1,2}(B_{1,2}^{a_1,a_2,a_3})$ is $\{5, 6, 6, 8, 10, 11, 15\}$.

4.2 Hardness of Naive Computation

A feature of shill bids is that the scheming agent can increase his/her utility when he/she bids the evaluation values divided into multiple bids. We can determine the possibility of shill bidders by comparing the utilities between a calculation using an individual agent's evaluation value and one using the merged values. The number of merged investments is calculated as $2^n - n - 1$, namely the computational cost is $O(2^n)$. Furthermore, when we determine the payments for winners in VCG, we need a computational cost of $O(2^n)$ for each winner. Therefore, a huge computational cost is required to judge whether an allocation is shill-biddable. Consequently, we propose a cost-cutting method for finding shill-biddable allocations and a heuristic method for auctions involving a massive number of bidders.

5 Brute Force Algorithm

5.1 Shill-bidders must be winners

Agents who create shill bidders basically cannot increase their utilities without the shill bidder winning in an auction. We can show this feature through the following theorem.

Theorem 1 (Shill-bidders must be winners). *An agent who creates shill bidders can not increase his/her utility unless a shill bidder wins.*

We prove that bidder agent i's utility u_i does not decrease when the agent's shill bidder does not win in an auction. When bidder i does not create any shill bidders, bidder i's payment p_i can be illustrated by using the following equation.

$$p_i = \sum_{i \neq j} v_j(G^*_{\sim i}) - \sum_{i \neq j} v_j(G^*).$$

When the bidder agent creates shill bidders, bidder i's payment p'_i is

$$p'_i = \sum_{i \neq j} v_j(G'^*_{\sim i}) - \sum_{i \neq j} v_j(G'^*).$$

Here, we show $p'_i \geq p_i$ in the following proof. We assume bidder i's shill bidders do not win in the auction. Also, the set of allocations in the auction does not change. Namely, $G' = G \not\ni s$ holds. The difference between p'_i and p_i is shown as follows.

$$p'_i - p_i = \sum_{i \neq j} v_j(G'_{\sim i}) - \sum_{i \neq j} v_j(G') - (\sum_{i \neq j} v_j(G_{\sim i}) - \sum_{i \neq j} v_j(G))$$
$$= \sum_{i \neq j} v_j(G'_{\sim i}) - \sum_{i \neq j} v_j(G_{\sim i}).$$

Next, we show $\sum_{i \neq j} v_j(G'_{\sim i}) \geq \sum_{i \neq j} v_j(G_{\sim i})$.

We assume that the number of items is m, where these include an a item's set $M = \{a_1, a_2, \ldots, a_m\}$. A set of bids (bundles) is assumed to be

$$B = \{B_1, \ldots, B_{i-1}, B_{i+1}, \ldots, B_n\} \in G_{\sim i}.$$

However, the set of bids is shown as

$$B' = B \cup \{B'_1, \ldots, B'_{n'}\} \in G'_{\sim i}.$$

The set $\{B'_1, \ldots, B'_{n'}\}$ is the subset of the shill bid.

The objective function of the winner determination problem solved over a strictly larger set of bids (that is, one with additional shill bids) cannot decrease. By adding bids the winner determination value never goes down. Thus, shill bidders' utilities do not increase when the shill bidder does not win.

5.2 Winner-based algorithm

Based on the above theorem1, we propose a method to determine the possibility of shill bidding by using the winners' evaluation values in a combinatorial auction. In this paper, we assume the case where there is one agent who is engaging in shill bidding, because the analysis is complicated when there are multiple agents who create shill bidders. Shill bidders are created by separating the bids of an original agent. Namely, the original agent can increase the utility by dividing his/her bidding actions. When the original bidder and his/her shill bidders are winners of an auction, we can determine how much the agent increases the utility by comparing the utilities of divided bidding and merged investment bidding. Intuitively, we propose a winner-based algorithm, which is based on a comparison of utilities between bidding in real auctions and bidding based on our method. We describe our proposed algorithm in detail.

Input: evaluation values of bundles for each player.
Output: True if there is a shill bid.
False if there is no shill bid.

Function Detecting a Shill bid
begin
Determining winners and calculating
payments based on GVA.
Creating a power set S for a set of players.

for each $s \in S$
  Merging players' evaluation values in s
    by merge function $f(s)$.
  Determining winners and calculating payments
    based on merged evaluation values by GVA.
  $u_{f(s)}$:= the utility of s after merged.
  u_{sum_s} := sum of the utilities in s before merged.
  if $u_{f(s)} < u_{sum_s}$
    return True
  return False
end.

First, bidders bid their evaluation values based on a set of bundled items. Next, the items' allocation, that is, the combination where social surplus is maximized, is computed based on the GVA's method. Namely, winners are selected. Then, each agent's utility is calculated. Here, our algorithm is employed in this phase. First, all agents' evaluation values are merged, and the total surplus is re-calculated using the merged evaluation value. Second, the agent's utility in merged investment is compared with the sum of the agents' utilities in divided bidding.

Here, we define a merge function f and show the merged agents' evaluation values and payment from a set of agents. We assume that the set of agents who are merged is $\{i, i+1, \ldots, j\} \in N$.

Definition 2 (Merge function). *f is the merge function, which is defined as $f(i, i+1, \ldots, j)$ when a set of agents $\{i, i+1, \ldots, j\}$ are merged.*

To compare bidder's utilities between the case in the real auction and the case based on merged investment, merged agents' evaluation values v, payments p, and utilities u is defined as follows.

$v_{f(i,i+1,\ldots,j)}$ is merged evaluation values based on agents $\{i, i+1, \ldots, j\}$s' evaluation values . $p_{f(i,i+1,\ldots,j)}$ is agents $\{i, i+1, \ldots, j\}$s' payment amount under merged investment.

$u_{f(i,i+1,\ldots,j)}$ is agents $\{i, i+1, \ldots, j\}$s' utilities under merged investment.

In this section, we assume that $v_{f(i,i+1,\ldots,j)}$ is the following maximum function among agents' evaluation values for each item.

Maximum selection method The maximum selection method is shown that $v_{f(i,i+1,\ldots,j)}$ is shown as $(\max_{i,i+1,\ldots,j}\{v_i^{a_1}\}, \max_{i,i+1,\ldots,j}\{v_i^{a_2}\}, \ldots, \max_{i,i+1,\ldots,j}\{v_i(B_i^{a_1,\ldots,a_m}\})$ for agent i's evaluation value $(v_i^{a_1}, v_i^{a_2}, \ldots, v_i(B_i^{a_1,\ldots,a_m}))$.

If there is the possibility of shill bidders, the following equation holds.

$$\sum_{i,i+1,\ldots,j} u_{i,i+1,\ldots,j} > u_{f(i,i+1,\ldots,j)}$$

$\sum_{i,i+1,\ldots,j} u_{i,i+1,\ldots,j}$ is the total sum of the agents' $\{i, i+1, \ldots, j\}$ utilities in the case of divided bidding in the allocation of the same items.

Here, we show an example of merge valuations by using the maximum selection function. For example, assume there are four agents and three items $M = (a_1, a_2, a_3)$ in an auction. Each agent bids for a bundle, that is, $\{(a_1), (a_2), (a_3), (a_1, a_2), (a_1, a_3), (a_2, a_3), (a_1, a_2, a_3)\}$.

Agent 1's value $v_1(B_1^{a_1,a_2,a_3}) : \{7, 0, 0, 7, 7, 0, 7\}$
Agent 2's value $v_2(B_2^{a_1,a_2,a_3}) : \{0, 0, 0, 0, 0, 0, 16\}$
Agent 3's value $v_3(B_3^{a_1,a_2,a_3}) : \{0, 6, 0, 6, 0, 6, 6\}$
Agent 4's value $v_4(B_4^{a_1,a_2,a_3}) : \{0, 0, 8, 0, 8, 8, 8\}$

In GVA, winners are decided as the combination in which social surplus is maximum, that is, \$21 when agents 1, 3 and 4 are selected. Agent 1 can purchase item a_1 for \$2, agent 3 can purchase item a_3 for \$1, and agent 4 can purchase item a_4 for \$3. Each agent's utility is calculated for \$5.

Here, to detect a shill bidder, our algorithm merges the agents' evaluation values. When agent 1's and agent 3's evaluation values are merged, the merged value $v_{f(1,3)}$ is shown as $\{7, 6, 0, 7, 7, 6, 7\}$.

The merged payment of agents 1 and 3, $p_{f(1,3)}$, is \$8, which is calculated as $16 - 8 = 8$. Agent 4's payment p_4 is \$3. $u_{f(1,3)}$ is calculated as $13 - 8 = 5$. This shows that agent 1 and 3 can increase their utilities by dividing their evaluation values if agents 1 and 3 are identical. Namely, this situation indicates the possibility of a shill bidder.

The combination of merged evaluation values is $\{v_{f(1,2)}, v_{f(1,3)}, v_{f(1,4)}, v_{f(2,3)}, v_{f(2,4)}, v_{f(3,4)}, v_{f(1,2,3)}, v_{f(1,2,4)}, v_{f(1,3,4)}, v_{f(2,3,4)}, v_{f(1,2,3,4)}\}$. When all possible outcomes are enumerated, we can determine which agent might be a shill bidder. However, the combinations of merged investment increase exponentially when the number of agents and items increase. To solve this problem, we propose a greedy algorithm in the next section.

6 Handling a Massive Number of Bidders

The combination of merged evaluation values can be computed as $\sum_{l=2}^{m} {}_mC_l$. For example, when the number of agents is 10, the combination is calculated to be $\sum_{l=2}^{10} {}_{10}C_l = 1003$. Naturally, winner determination using GVA is an NP-hard problem. Namely, in the above example, the calculation of GVA's method is conducted 1003 times.

Here, to solve the computational cost problem in a massive number of bidders, we propose an algorithm to find shill bidders. We assume the following condition.

Assumption 4 (Possibility of shill bids) *When one of the bidders' evaluation values involves a bundle's evaluation value that is more than the sum of all items' evaluation values, shill bids can be successful. Namely,* $v_i(B_i^{k,k+1,\ldots,l}) \leq \sum_k^l v_i^k$.

We show the algorithm as follows. The feature of our algorithm is searching for agents who might be shill bidders. To decrease the search space, our algorithm

searches for agents, who might be shill bidders, based on pruning of searching candidate.
[Algorithm]

(**Step 1**) Winners are determined based on bidders' evaluation values. The winners' utilities u_i are reserved.

(**Step 2**) Our system reserves winners' evaluation values and agents' evaluation values, which determine the winners' payments.

(**Step 3**) Evaluation values as described in the above assumption are searched for. Based on the search, the set of evaluation values is judged to fall into one of the following two cases.
(1) The type of evaluation value shown in the above assumption does not exist.
(2) The type of evaluation value shown in the above assumption does exist.

(**Step 4**) In the former case (1), winners' payment amounts are calculated. In the latter case (2), the process moves to (**step5**).

(**Step 5**) Our system finds bidders whose payments are determined based on the evaluation value of an agent who bids according to the above assumption.

(**Step 6**) Winners' evaluation values are merged based on the method shown in Section 4. Winners' utilities $u_{f(\cdot)}$ in each merged investment are reserved. Agents' utilities u_i are compared with the merged agents' utilities $u_{f(\cdot)}$.

(**Step 7**) Based on the comparison made in (**step 5**), when there are cases in which the difference between u_i and $u_{f(\cdot)}$ is not equal, a list of agents who take part in these cases is shown to the auction operator.

7 Experiments

We conducted a experiment to show the efficiency of the winner-based algorithm shown in Section 4.2. In the experiment, we measured the average elapsed time to judge whether the given bids can include shill bids or not. Table 1 shows a result where the number of items is 3. We varied the number of bidders from 3 to 17. If there is no possibility of shill bids, for simplicity, the elapsed time is defined as the elapsed time to search all combinations of bidders. We created 1,000 different problems for each number of bidders.

The evaluation values for player i is determined as follows: First, the evaluation values for each single item are determined for each bidder based on uniform distribution. Second, we determined whether items in a bundle is substitute or compliment at the probability of 0.5. Third, if the items in a bundle are compliment, the evaluation

Table 1. Result of the experiments

Num of bidders	Brute Force	Winner-based
3	1.2	1.07
4	2.9	1.13
5	3.97	1.36
6	6.76	2.01
7	11.79	1.68
8	14.12	1.6
9	52.15	1.83
10	79.72	2.18
11	145.69	2.38
12	362.98	2.57
13	834.61	2.69
14	1489.49	2.68
15	1397.02	2.49
16	5906.26	3.01
17	10096.04	3.13

value of the bundle is defined as sum of evaluation values for the items in the bundle. If the items are substitute, the evaluation value of the bundle is defined as maximum of evaluation values for the items in the bundle.

In the both algorithms, we need to compute combinatorial optimization problems in GVA. Thus, in this experimentation, we utilize the BIN and the other improving methods in the CASS algorithm [2]. The experimental environment is Java SDK 1.4.2, Mac OS X 10.3, Power PC G5 2Ghz dual, and 1.5 GB memory.

In terms of the heuristic algorithm shown in Section 5, the computation cost is clearly similar to or less than that of the winner-based algorithm. Thus, we focus on the winner-based algorithm shown in Section 4.2.

In the worst case, the brute force algorithm(BF) clearly needs an exponential time to judge whether a shill bid is included or not. Thus, in the table 1, the elapsed time of the brute force algorithm increased exponentially. On the contrary, the winner-based algorithm we proposed needs a linear time. The reason of the efficiency of the winner-based algorithm can be described as follows: Even the number of bidders increases, the number of winners must be lower than the number of items. Thus, in the winner-based algorithm, we do not need exponential number of combinations of bidders. The number of combinations are always lower then the number of items.

8 Conclusions

This paper proposed a method for detecting shill bids in combinatorial auctions. Our algorithm can judge whether there might be a shill bid from the results of GVA's procedure. However, a straightforward way to detect shill bids requires an exponential amount of computing power because we need to check all possible combinations of bidders. Therefore, in this paper we proposed an improved method for finding shill bidders. The method is based on winning bidders. The results demonstrated that the proposed method succeeds in reducing the computational cost of finding shill bids.

In actual auctions, it is important for sellers and an auctioneer to know whether shill bidders include in an auction or not. If the auctioneer can know that shills include in bidders, the auctioneer can close the bidding to avoid reducing sellers' earnings. Furthermore, it is useful for the auctioneer to know who are shill bidders. The auctioneer can protect and secure from shills who are removed in subsequent auctions. Moreover, in the developed auction system such as an automatic bidding system and an intelligent auction system, if the system detects shill bidders based on the bid value data sets, auctioneers can shut out malicious users from auction sites.

References

1. E. H. Clarke. Multipart pricing of public goods. *Public Choice*, 11:17–33, 1971.
2. Y. Fujishima, K. Leyton-Brown, and Y. Shoham. Taming the computational complexity of combinatorial auctions: Optimal and approximate approaches. In *Proc. of the 16th International Joint Conference on Artificial Intelligence (IJCAI99)*, pages 548–553, 1999.
3. T. Groves. Incentives in teams. *Econometrica*, 41:617–631, 1973.
4. D. Lehmann, L. I. O'Callaghan, and Y. Shoham. Truth revelation in approximately efficient combinatorial auctions. *Journal of the ACM*, 49(5):577–602, 2002.
5. K. Leyton-Brown, M. Tennenholtz, and Y. Shoham. An algorithm for multi-unit combinatorial auctions. In *Proc. of 17th National Conference on Artificial Intelligence (AAAI2000)*, 2000.
6. P. Milgrom. *Putting Auction Theory to Work*. Cambridge University Press, 2004.
7. T. Sandholm. An algorithm for optimal winnr determination in combinatorial auctions. In *Proc. of the 16th International Joint Conference on Artificial Intelligence(IJCAI'99)*, pages 542–547, 1999.
8. T. Sandholm, S. Suri, A. Gilpin, and D. Levine. Winner determination in combinatorial auction generalizations. In *Proc. of the 1st International Joint Conference on Autonomous Agents and Multi-Agent Systems (AAMAS02)*, pages 69–76, 2002.
9. W. Vickrey. Counterspeculation, auctions, and competitive sealed tenders. *Journal of Finance*, XVI:8–37, 1961.
10. M. Yokoo, Y. Sakurai, and S. Matsubara. Bundle design in robust combinatorial auction protocol against false-name bids. In *Proceedings of the 17th International Joint Conference on Artificial Intelligence (IJCAI-2001)*, pages 1095–1101, 2001.
11. M. Yokoo, Y. Sakurai, and S. Matsubara. Robust multi-unit auction protocol against false-name bids. In *Proceedings of the 17th International Joint Conference on Artificial Intelligence (IJCAI-2001)*, pages 1089–1094, 2001.
12. M. Yokoo, Y. Sakurai, and S. Matsubara. The effect of false-name bids in combinatorial auctions: New fraud in Internet auctions. *Games and Economic Behavior*, 46(1):174–188, 2004.

The Prediction of Partners' Behaviors in Self-Interested Agents

Fenghui Ren and Minjie Zhang

School of Computer Science and Software Engineering
University of Wollongong, Australia
{fr510, minjie}@uow.edu.au

Summary. Prediction partners' behaviors in negotiation has been an active research direction in recent years. By employing the estimation results, agents can modify their negotiation strategies in order to achieve an agreement much quicker or to gain higher benefits. Some of estimation strategies have been proposed by researchers, and most of them are based on machine learning mechanisms. However the machine learning based approach may not work well in some open and dynamic domains for the reasons of (1) lacking of sufficient data to train the system, and (2) requesting plenty of resources in each training process. Furthermore, because the estimation results may have errors, so single result maybe not accurate and practical enough in most situations. In order to address these issues mentioned above, we propose a quadratic regression analysis approach to predict partners' behaviors in this paper. The proposed approach is based only on the history of the offers during the current negotiation and does not require any training process in advance. This approach can estimate an interval of behaviors according to an accuracy requirement. The experimental results illustrate that by employing the proposed mechanism, agents can gain more accurate estimation results on partners' behaviors by comparing with other two estimation functions.

1 Introduction

Negotiation is a means for agents to communicate and compromise to reach mutually beneficial agreements [1] [2]. However, in most situations, agents do not have complete information about their partners' negotiation strategies, and may have difficulties to make a decision on future negotiation, such as how to select suitable partners for further negotiation [3] [4] or how to generate a suitable offer in next negotiation cycle [5]. Therefore estimation approaches which can predict uncertain situations and possible changes in future are required for helping agents to generate good and efficient negotiation strategies. Research on partners' behaviors estimation has been a very active direction in recent years. Several estimation strategies are proposed [6] [7] [8] by researchers. However, as these estimation strategies are used in real applications, some of limitations are emerged.

Machine learning is one of the popular mechanisms adopted by researchers in agents' behaviors estimation. In general, this kind of approaches have two steps in

F. Ren and M. Zhang: *The Prediction of Partners' Behaviors in Self-Interested Agents*, Studies in Computational Intelligence (SCI) **110**, 157–170 (2008)
www.springerlink.com

order to estimate the agents' behaviors properly. In the first step, the proposed estimation function is required to be well trained by training data. Therefore, in a way, the performance of the estimation function is almost decided by the training result. In this step, data are employed as many as possible by designers to train a system. The training data could be both synthetic or collected from the real world. Usually, the synthetic data are helpful in training a function to enhance its problem solving skill for some particular issues, while the real world collected data can help the function to improve its ability in complex problem solving. After the system being trained, the second step is to employ the estimation function to predict partners' behaviors in future. However, no matter what and how many data are employed by designers to train the proposed function, it is unsuspicious to say that the training data will be never comprehensive enough to cover all situations in reality. Therefore, even though an estimation function is well trained, it is also very possible that some estimation results do not make sense at all for some kind of agents whose behaviors' records are not included in the training data. At present stage, as the negotiation environment becomes more open and dynamic, agents with different kinds of purposes, preferences and negotiation strategies can enter and leave the negotiation dynamically. So the machine learning based agents' behaviors estimation functions may not work well in some more flexible application domains for the reasons of (1) lacking of sufficient data to train the system, and (2) requesting plenty of resources in each training process.

Another insufficiency of present estimation functions is the representation of the predication results. Most of present estimation functions can only predict a single value but not an interval as result to represent agents' possible behaviors. However in agents behaviors estimation, the representation of predication results in the form of an interval is more reasonable than a single value because (1) an interval is more accuracy than a value to represent something uncertain; (2) an interval provides more flexible choices for agents to modify their negotiation strategies than a value; and (3) an interval can easily be further adopted by other AI mechanisms, such as Fuzzy Logic mechanism, to administrate agents' behaviors than a value. In order to increase both the accuracy and flexibility for an agent to estimate its partners' behaviors and to perform a reasonable response, the behavior estimation result should be represented in the form of an interval.

In order to address those issues mentioned above, in this paper we propose a quadratic regression approach to analyze and estimate partners' behaviors in negotiation. According to our knowledge, this is the first time that the regression analysis approach is employed to estimate partners' behaviors in negotiation. By comparing with machine learning mechanisms, the proposed approach only uses the historical offers in the current negotiation to estimate partners' behaviors in future negotiation and does not require any additional training process. So the proposed approach is very suitable to work under an open and dynamic negotiation environment, and to make a credible judgements on partners' behaviors timely. Also, because the proposed approach does not make any assumption on agents' purposes, preferences and negotiation strategies, it can be employed widely in negotiation by different types of agents. Furthermore, the proposed approach not only represents the estimation results in the form of interval, but also gives the probability that each individual situation may happen in future. By employing the proposed representation format, agents can have an overview on partners' possible behaviors and their favoritism easily, and then modify their own strategies based on these distributions. Therefore, the proposed approach provides more flexible choices to agents when they make decision in negotiation.

The rest paper is organized as follows: Section 2 introduces both the background and assumption of our proposed approach; Section 3 introduces the way that how the proposed regression analysis works; Section 4 introduces the ways to predict partners' behaviors under different accuracy requirements based on the regression analysis results; Section 5 illustrates the performance of the proposed quadratic regression function through experiments and advantages of the proposed function by comparing with other two estimation approaches; and Section 7 concludes this paper and outlines future works.

2 Background

In this section, we introduce the background about the proposed quadratic regression function for partners' behaviors estimation in negotiation. The regression analysis is a combination of mathematics and probability theory which can estimate the strength of a modeled relationship between one or more dependent variables and independent variables. In this paper, we propose a quadratic regression function to predict the partners' behaviors. The proposed quadratic regression function is given as follows:

$$u = a \times t^2 + b \times t + c \tag{1}$$

where u is the expected utility gained from a partner, t $(0 \leq t \leq \tau)$ is the negotiation cycle and a, b and c are parameters which are independent on t. Based on this proposed quadratic regression function, it is noticed that four types of partners' behaviors [9] are distinguished.

- $a > 0$: the rate of change in the slope is increasing, corresponding to smaller concession in the early cycles but large concession in later cycles.
- $a = 0$ and $b \neq 0$: the rate of change in the slope is zero, corresponding to making constant concession throughout the negotiation.
- $a < 0$: the rate of change in the slope is decreasing, corresponding to large concession in early cycle but smaller concession in later cycles.
- $a = 0$ and $b = 0$: The rate of change of the slope and the slope itself are always zero, corresponding to not making any concession throughout the entire negotiation.

In the following Section, we will introduce the proposed quadratic regression function to analysis and estimate partners' possible behaviors in a more efficient and accurate way.

3 Regression Analysis on Partners' Behaviors

In order to simplify the problem, we firstly transfer the proposed quadratic function 1 to a linear function as follows. Let

$$\begin{cases} x = t^2 \\ y = t \end{cases} \tag{2}$$

Then Equation 2 can be rewritten as follows:

$$u = a \times x + b \times y + c \tag{3}$$

where both a and b are independent on variable x and y. Let pairs $(t_0, \hat{u_0}), \ldots, (t_n, \hat{u_n})$ are the instances from each negotiation cycle according to the time order. Then the distance (ε) between the real utility value ($\hat{u_i}$) and the expected value (u_i) should obey the Gaussian distribution which is $\varepsilon \sim N(0, \sigma^2)$, where $\varepsilon = \hat{u_i} - a \times x_i - b \times y_i - c$. Because for each $\hat{u_i} = a \times x_i + b \times y_i + c + \varepsilon_i$, $\varepsilon_i \sim N(0, \sigma^2)$, $\hat{u_i}$ is distinctive, then the joint probability density function for $\hat{u_i}$ is:

$$\begin{aligned} L &= \prod_{i=1}^{n} \frac{1}{\sigma\sqrt{2\pi}} \exp[-\frac{1}{2\sigma^2}(\hat{u_i} - ax_i - by_i - c)^2] \\ &= (\frac{1}{\sigma\sqrt{2\pi}})^n \exp[-\frac{1}{2\sigma^2}\sum_{i=1}^{n}(\hat{u_i} - ax_i - by_i - c)^2] \end{aligned} \tag{4}$$

where L indicates the probability that a particular $\hat{u_i}$ may happen. Because each $\hat{u_i}$ comes from historical records, so we should keep their probabilities as L's maximum value. Obviously, in order to make L to achieve its maximum, $\sum_{i=1}^{n}(\hat{u_i} - ax_i - by_i - c)^2$ should achieve its minimum value. Let

$$Q(a,b,c) = \sum_{i=1}^{n}(\hat{u_i} - ax_i - by_i - c)^2 \tag{5}$$

We calculate the first-order partial derivative for $Q(a,b,c)$ on a, b and c respectively, and let their results equal to zero, which are shown as follows:

$$\begin{cases} \frac{\partial Q}{\partial a} = -2\sum_{i=1}^{n}(\hat{u_i} - ax_i - by_i - c)x_i = 0 \\ \frac{\partial Q}{\partial b} = -2\sum_{i=1}^{n}(\hat{u_i} - ax_i - by_i - c)y_i = 0 \\ \frac{\partial Q}{\partial c} = -2\sum_{i=1}^{n}(\hat{u_i} - ax_i - by_i - c) = 0 \end{cases} \tag{6}$$

then the above equations equal to:

$$\begin{cases} (\sum_{i=1}^{n} x_i^2)a + (\sum_{i=1}^{n} x_i y_i)b + (\sum_{i=1}^{n} x_i)c = \sum_{i=1}^{n} x_i \hat{u_i} \\ (\sum_{i=1}^{n} x_i y_i)a + (\sum_{i=1}^{n} y_i^2)b + (\sum_{i=1}^{n} y_i)c = \sum_{i=1}^{n} y_i \hat{u_i} \\ (\sum_{i=1}^{n} x_i)a + (\sum_{i=1}^{n} y_i)b + nc = \sum_{i=1}^{n} \hat{u_i} \end{cases} \tag{7}$$

Let PU, PA, PB and PC are the coefficient matrices as follows:

$$PU = \begin{vmatrix} \sum_{i=1}^{n} x_i^2 & \sum_{i=1}^{n} x_i y_i & \sum_{i=1}^{n} x_i \\ \sum_{i=1}^{n} x_i y_i & \sum_{i=1}^{n} y_i^2 & \sum_{i=1}^{n} y_i \\ \sum_{i=1}^{n} x_i & \sum_{i=1}^{n} y_i & n \end{vmatrix} \tag{8}$$

$$PA = \begin{vmatrix} \sum_{i=1}^{n} x_i \hat{u}_i & \sum_{i=1}^{n} x_i y_i & \sum_{i=1}^{n} x_i \\ \sum_{i=1}^{n} y_i \hat{u}_i & \sum_{i=1}^{n} y_i^2 & \sum_{i=1}^{n} y_i \\ \sum_{i=1}^{n} \hat{u}_i & \sum_{i=1}^{n} y_i & n \end{vmatrix} \tag{9}$$

$$PB = \begin{vmatrix} \sum_{i=1}^{n} x_i^2 & \sum_{i=1}^{n} x_i \hat{u}_i & \sum_{i=1}^{n} x_i \\ \sum_{i=1}^{n} x_i y_i & \sum_{i=1}^{n} y_i \hat{u}_i & \sum_{i=1}^{n} y_i \\ \sum_{i=1}^{n} x_i & \sum_{i=1}^{n} \hat{u}_i & n \end{vmatrix} \tag{10}$$

$$PC = \begin{vmatrix} \sum_{i=1}^{n} x_i^2 & \sum_{i=1}^{n} x_i y_i & \sum_{i=1}^{n} x_i \hat{u}_i \\ \sum_{i=1}^{n} x_i y_i & \sum_{i=1}^{n} y_i^2 & \sum_{i=1}^{n} y_i \hat{u}_i \\ \sum_{i=1}^{n} x_i & \sum_{i=1}^{n} y_i & \sum_{i=1}^{n} \hat{u}_i \end{vmatrix} \tag{11}$$

Because $PU \neq 0$, the parameters a, b and c have an unique solution, which is

$$\begin{cases} a = \frac{PA}{PU} \\ b = \frac{PB}{PU} \\ c = \frac{PC}{PU} \end{cases} \tag{12}$$

By adopting the parameters above, we define a quadratic function which can be employed to predict agents' behaviors.

4 Partners' Behaviors Prediction

In last section, we proposed a quadratic regression function to predict partners' behaviors, and also specified how to decide parameters a, b and c. However, it has to be mentioned that the proposed quadratic regression function can only provide an estimation on partners' possible behaviors, which might not exactly accord with the partners' real behaviors. In reality, the real behaviors should close to the estimation behaviors, and the more closer to the estimated behaviors, the higher probability it may happen. So we can deem that the differences (ε) between the estimation behaviors and the real behaviors obey the Gaussian distribution $N(\varepsilon, \sigma^2)$. Thus, if the deviation σ^2 can be calculated, we can make a precise decision on the range of partners' behaviors. It is known that there is more than 68% probability that partners' expected behaviors locate in the interval $[u - \sigma, u + \sigma]$, more than 95% that partners' expected behaviors in $[u - 2\sigma, u + 2\sigma]$, and more than 99% in the interval $[u - 3\sigma, u + 3\sigma]$. In this section, we introduce the proposed way to calculate the deviation σ and to estimate the interval for partners' behaviors.

In order to calculate the deviation σ, we firstly calculate the distance between the estimation results (u_i) and the real results on partners' behaviors ($\hat{u}_i$) as follows:

$$d_i = \hat{u}_i - u_i \tag{13}$$

It is known that all d_i ($i \in [1, n]$) obey the Gaussian distribution $N(0, \sigma^2)$. Then σ can be calculated as follows:

$$\sigma = \sqrt{\frac{\sum_{i=1}^{n}(d_i - \overline{d})^2}{n}} \tag{14}$$

where,

$$\overline{d} = \frac{1}{n}\sum_{i=1}^{n} d_i \tag{15}$$

Now we can approve that the partners' behaviors, which obey Gaussian distribution $N(u, \sigma^2)$. By employing the Chebyshev's inequality [10], we can calculate (1) the interval of partners' behaviors according to any accuracy requirements; and (2) the probability that any particular behavior may happen on potential partners in future.

The Chebyshev's inequality is given as follows:

$$P(|X - \mu| \geq \varepsilon) \leq \frac{\sigma^2}{\varepsilon^2} \tag{16}$$

where X is an instance, μ is the mathematical expectation, σ is the deviation and ε is the accuracy requirement. $|X - \mu| \geq \varepsilon$ represents the interval of partners' behaviors and $\frac{\sigma^2}{\varepsilon^2}$ is the probability that the behavior X will happen in future.

5 Experiments

In this section, we demonstrate several experiments to test our proposed regression analysis approach. We display the prediction results by using the proposed approach in each negotiation cycle. Also, we compare the proposed quadratic regression approach with the Tit-For-Tat approach [9] and random approach. The experimental results illustrate the outstanding performance of our proposed approach.

In order to simplify the implement process, all agents in our experiment employ the NDF [11] negotiation strategy. The partners' behaviors cover all possible situations in reality, which are conceder, linear and boulware. In experiments, we use the average error (EA) to evaluate the experimental results. Let u_i be the predict result in cycle i and $\hat{u}_i$ be the real instance in cycle i, then the AE_i is defined as follows:

$$AE_i = \frac{\sum_{k=1}^{i}|\hat{u}_i - u_i|}{i} \tag{17}$$

The AE_i indicates the difference between the estimated results and the real value. The smaller the value of AE_i, the better the prediction result.

5.1 Scenario 1

In the first scenario, a buyer want to purchase a mouse pad from a seller. The acceptable price for the buyer is in $[\$0, \$1.4]$. The deadline for buyer to finish this purchasing process is 11 cycles. In this experiment, the buyer adopts conceder behavior in the negotiation, and the seller employs the proposed approach to estimate the buyer's possible price in the next negotiation cycle. The estimated results are displayed in Figure 1 and the regression function is:

$$u = -0.002 * t^2 + 0.055 * t + 0.948$$

It can be seen that in the $8th$ negotiation cycle, according to instances, the proposed approach estimates a price of $\$1.26$ from the buyer in next cycle. Then according to the historical record in the $8th$ cycle, the real price given by the buyer in this cycle is $\$1.26$ which is exactly same as the estimation price. Furthermore, it can be seen that in cycle 4, 6, 9 and 10, the estimated prices are also same as the real value. The estimation prices for $2th$, $3th$ and $7th$ cycles are $\$1.05$, $\$1.10$ and $\$1.25$ respectively, and the real prices given by buyer in these cycles are $\$1.07$, $\$1.13$, and 1.26, which only have very little difference between the estimated prices and real prices in these cycles. According to Figure 1, all real prices are located in the interval of $[\mu - 2\sigma, \mu + 2\sigma]$, where μ is the estimated price and σ is the changing span. The $AE_{10} = 0.015$, which is only 1% of buyer's reserve price. Therefore, the prediction results by employing the proposed approach are very reliable.

In Figure 2, we illustrate the comparison results between the proposed approach and other two estimation approaches (Tit-For-Tat and random approach). It can be seen that even though the Tit-For-Tat approach can follow the trend of changes in buyer's price, the $AE_{10} = 0.078$ which is five times of our proposed approach. For the random approach, it even cannot catch the main trend. The AE_{10} for the random approach is 0.11, which is ten times as much as our proposed approach. The experimental results convince us that the proposed approach outperforms both Tit-For-Tat and random approaches very much when a buyer adopts conceder negotiation behavior.

5.2 Scenario 2

In the second scenario, a buyer wants to buy a keyboard from a seller. The desired price for the buyer is in the interval of $[\$0, \$14]$. We let the buyer to employ the linear negotiation strategy, and still set the deadline to 11 cycles. The seller will employ our proposed prediction function to estimate the buyer's offer. The estimated results are illustrated in Figure 3 and the estimated quadratic regression function is:

$$u = -0.015 * t^2 + 1.178 * t - 0.439$$

It can be seen that in the $3th$, $5th$ and $8th$ cycles, the estimated prices are exactly same as the real offers given by the buyer. The biggest difference between the estimated price and the real value is just 0.4, which happens in the $9th$ cycle. The

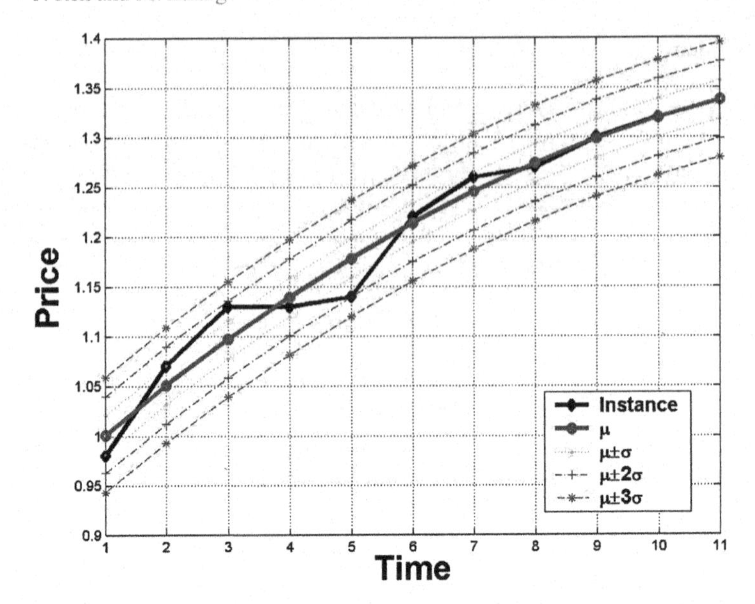

Fig. 1. Prediction results for scenario 1

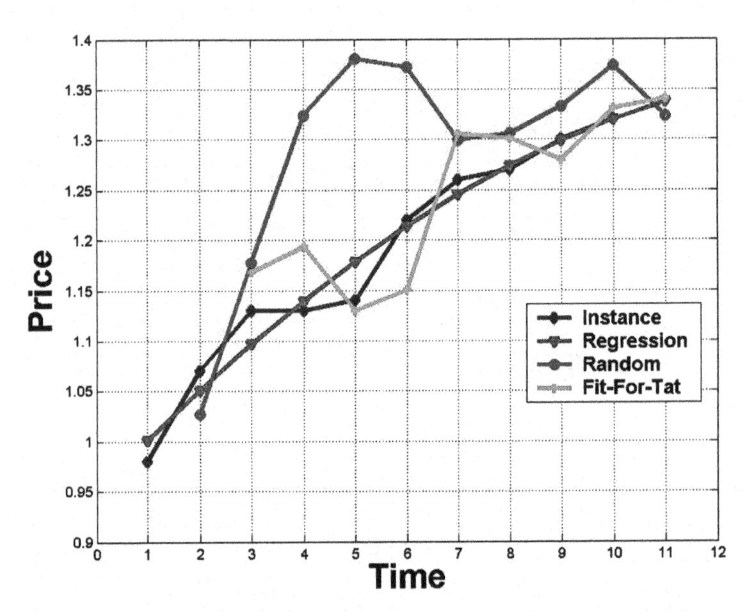

Fig. 2. Prediction results comparison for scenario 1

average error in this experiment is only $AE_{10} = 0.24$, which is no more the 2% of the buyer's reserve price. The estimated quadratic regression function fits the real prices very well.

In Figure 4, the comparison results among Tit-For-Tat approach, random approach and our proposed approach are illustrated. It can be seen that the proposed approach is much more close the real price than the other two approaches. The average error for the Tit-For-Tat approach is $AE_{10} = 2.52$, which is 18% of the buyer's reserve price. The average error for the random approach is very high, which is $AE_{10} = 4.82$ and 34% of the buyer's reserve price. The second experimental results demonstrate that when partners perform as the linear behaviors, the proposed approach also outperforms other two approaches.

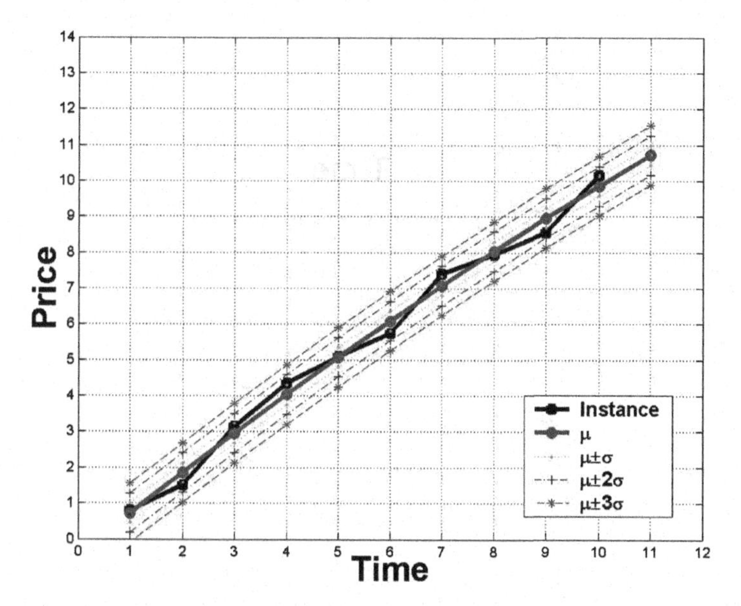

Fig. 3. Prediction results for scenario 2

5.3 Scenario 3

In the third scenario, a buyer wants to purchase a monitor from a seller. The suitable price for the buyer is in [\$0, \$250]. In this experiment, the buyer employs a boulware strategy in the negotiation. The deadline is still 11 cycles. The estimated quadratic function is:

$$u = 3.038 * t^2 - 12.568 * t + 15.632$$

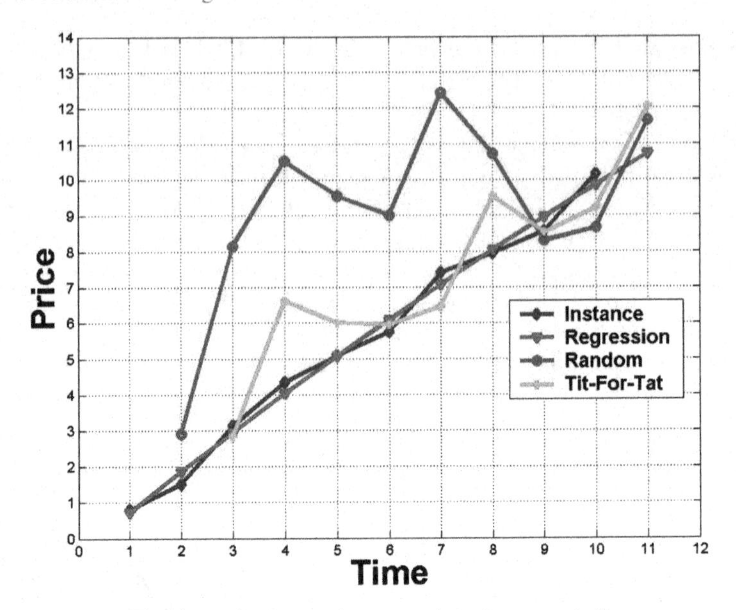

Fig. 4. Prediction results comparison for scenario 2

The estimated results are shown in Figure 5, it can be seen that the proposed quadratic regression approach predicted buyer's prices successfully and accurately. Except the $4th$ and $8th$ cycles, others estimated prices almost have no difference with the buyer's real offers. The average error in this experiment is only $AE_{10} = 4.07$, which is only 1.6% of the buyer's reserver price. Therefore, it is confident to say that from this estimation results, the seller can make very accurate judgement on the buyer's negotiation strategy, and make very reasonable responses in order to maximize its own benefit.

Finally, the Figure 6 illustrates the comparison results with other two estimation functions for the same scenario. For the Tit-For-Tat approach, the average error is $AE_{10} = 57.74$, which is 23% of the buyer's reserve price. For the random approach, the average error is $AE_{10} = 83.12$, which is 33% of the buyer's reserve price. Therefore, it can be seen that when the agent performs a boulware behavior, the proposed approach outperforms other two approaches very much.

From these experimental results in the above, we can say that the estimated quadratic function regression approach can estimate partners' potential behaviors successfully, and also the estimation results are accurate and reasonable enough to be adopted by agents to modify their strategies in negotiation. The comparison results among three types of agents' behaviors estimation approaches also demonstrate the outstanding performance of our proposed approach.

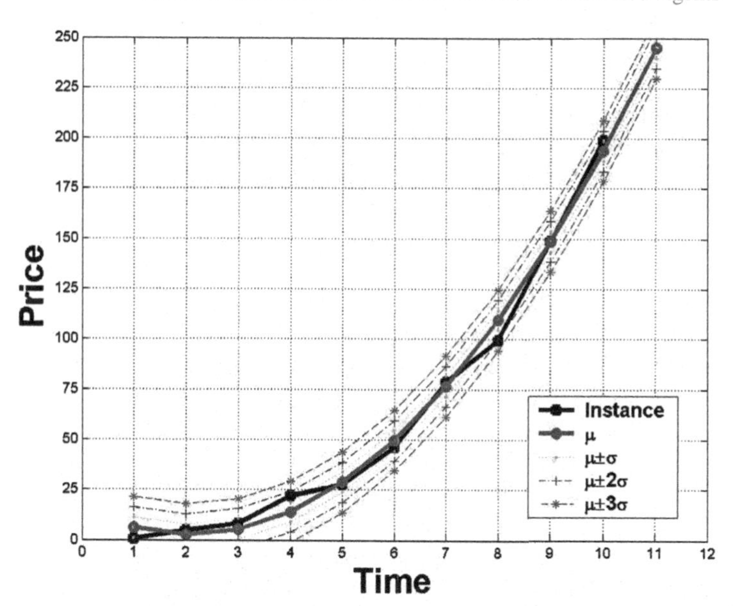

Fig. 5. Prediction results for scenario 3

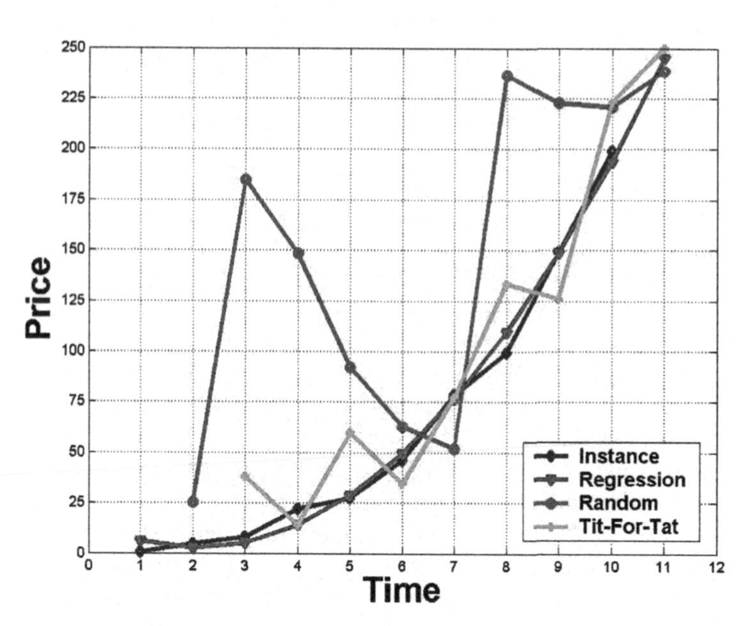

Fig. 6. Prediction results comparison for scenario 3

6 Related Work

In this section, we introduce some related works and give discussions on the proposed approach. In [12], Schapire et. al. proposed a machine learning approach based on a boosting algorithm. In the first place, the estimation problem is reduced to a classification problem. All training data are arranged in ascending order and then partitioned into groups equally. For each of the breakpoints, a learning algorithm is employed to estimate the probability that a new bid at least should be greater than the breakpoint. The final result of this learning approach is a function which gives minimal error rate between the estimated bid and the real one. Based on this function, agents' behaviors can be estimated. However, the accuracy of this approach is limited by the training data and classification approach. So applications based on this approach can hardly achieve a satisfactory level when negotiations happen in an open and dynamic environment.

In [13], Gal and Pfeffer presented another machine learning approach based on a statistical method. The proposed approach is trained by agents' behaviors according to their types firstly. Then for an unknown agent, it will be classified into a known kind of agents according to their similarities. Finally, based on these probabilities, the unknown agent's behavior is estimated by combining all known agents' behaviors. The limitation of this approach is that, in reality, it is impossible to train a system with all different types of agents. Therefore if an unknown agent belongs to a type which is excluded in the system, the estimation result may not reach an acceptable accuracy level.

Chajewska et. al. [8] proposed a decision-tree approach to learn and estimate agent's utility function. The authors assumed that each agent is rational which looks for maximum expected utility in negotiation. Firstly, a decision tree is established which contains all possible endings for the negotiation. Each possible ending is assigned with a particular utility value and possibility. Based on the partner's previous decisions on the decision tree, a linear function can be generated to analogy the partner's utility function, and each item in the function comes from an internal node on the decision tree. The limitation of this approach is the requirement that all possible negotiation endings and the corresponding probabilities should be estimated in advance, which is impossible in some application domains when the variance of negotiation issues is discrete or the negotiation environment is open and dynamic.

Brzostowski and Kowalczyk [14] presented a way to estimate partners' behaviors based only on the historical offers in the current negotiation. In this first place, partners' types are estimated based on the given functions. For each type of agents, a distinct prediction function is given to estimate agents' behaviors. Therefore, based on the classification about partners' types and their individual estimation functions, the proposed approach can predict partners' behaviors in next negotiation cycle. However, a partner can only perform as a time-dependent agent or a behavior-dependent agent, which limits some applications. Also the accuracy of classification on partners' types may impact the accuracy of prediction result.

By comparing our approach with the above estimation strategies on agents' behaviors, our proposed approach has two attractive merits. (1) The proposed approach do not need any training or preparation in advance, and it can estimate partners' behaviors based only on the current historical records and generate reasonable and accurate estimation results quickly and timely. Therefore, agents can save both space and time resources by employing the proposed approach; and (2) the proposed approach estimates partners' possible behaviors in the form of interval, and the probability that each particular behavior will happen in the future is also represented by the proposed quadratic regression function. Therefore, agents can adopt the estimation results by the proposed approach much easier and more convenient to administrate their own negotiation behaviors in future.

7 Conclusion and Future Work

In this paper, we proposed quadratic regression approach to estimate partners' behaviors in negotiation. We introduced the procedures to calculate the parameters in the regression function, and the method to predict partners possible behaviors. The experimental results demonstrate that the proposed approach is novel and valuable for the agents' behaviors estimation because (1) it is the first time that the regression analysis approach is applied on the agents' behaviors estimation; (2) the proposed approach does not need any training process in advance; (3) the representation format of the estimation results is easy to be further adopted by agents; and (4) the probability that each estimation behavior will happen in future on partners is also a significant criterion for agents to dominate their own behaviors in future.

The future works of this research will focus on two directions. Firstly, the multi-attribute negotiation is another promoting issue in recent years. Therefore, one of the emphases in our future works is to extend the proposed approach from the single-issue negotiation to the multi-issue negotiation. Secondly, as the negotiation environment becomes more open and dynamic, the proposed approach should be extended in order to predict not only agents' possible behaviors, but also impacts from potential changes on the negotiation environment.

References

1. S. Kraus. Strategic Negotiation in Multiagent Environments. The MIT Press, Cambridge, Massachusetts, 2001.
2. S. Fatima and M. Wooldridge and N. Jennings. Optimal Agendas for Multi-issue Negotiation, In: Second International Joint Conference on Autonomous Agents and Multi-Agent Systems (AAMAS03), pages 129–136. ACM Press, 2003.
3. J. Brzostowski and R. Kowalczyk. On Possibilistic Case-Based Reasoning for Selecting Partners in Multi-agent Negotiation. In: AI 2004: Advances in Artificial Intelligence, 17th Australian Joint Conference on Artificial Intelligence, volume 3339, pages 694–705. Cairns, Australia, Springer, Dec. 2004.
4. S. Munroe and M. Luck and M. d'Inverno. Motivation-Based Selection of Negotiation Partners. In: 3rd International Joint Conference on Autonomous Agents and Multiagent Systems (AAMAS 2004), pages 1520–1521. IEEE Computer Society, 2004.

5. S. Parsons and C. Sierra and N. Jennings. Agents that Reason and Negotiate by Arguing. Journal of Logic and Computation, 8(3): 261–292, June 1998
6. D. Zeng and K. Sycara. Bayesian Learning in Negotiation. International Journal of Human-Computer Studies, 48(1): 125–141, 1998.
7. R. Coehoorn and N.Jennings. Learning on Opponent's Preferences to Make Effective Multi-issue Negotiation Trade-offs. In: Proceedings of the 6th International Conference on Electronic Commerce, ICEC 2004, pages 59–68. Delft, Netherlands, ACM Press, October 2004.
8. U. Chajewska and D. Koller and D. Ormoneit. Learning An Agent's Utility Function by Observing Behavior. In: Proc. 18th International Conf. on Machine Learning, pages 35–42. Morgan Kaufmann, San Francisco, CA, 2001.
9. P. Faratin and C. Sierra and N. Jennings. Negotiation Decision Functions for Autonomous Agents. Journal of Robotics and Autonomous Systems, 24(3-4): 159–182, 1998.
10. Avaiable via wikipedia. http://en.wikipedia.org/wiki/Chebyshev's_inequality. Cited May 17 2007.
11. S. Fatima and M. Wooldridge and N. Jennings. An Agenda-Based Framework for Multi-Issue Negotiation. Artificial Intelligence, 152(1): 1–45, 2004
12. R. Schapire and P., D. McAllester and M. Littman and J. Csirik. Modeling Auction Price Uncertainty Using Boosting-based Conditional Density Estimation. In: Machine Learning, Proceedings of the Nineteenth International Conference (ICML 2002), pages 546–553. University of New South Wales, Sydney, Australia, Morgan Kaufmann, July 2002.
13. Y. Gal and A. Pfeffer. Predicting Peoples Bidding Behavior in Negotiation. In: 5th International Joint Conference on Autonomous Agents and Multiagent Systems (AAMAS 2006).
14. J. Brzostowski and R. Kowalczyk. Predicting partner's behaviour in agent negotiation. In: 5th International Joint Conference on Autonomous Agents and Multiagent Systems (AAMAS 2006).